Disclaimer

The publisher of this book is by no way associated with the National Institute of Standards and Technology (NIST). The NIST did not publish this book. It was published by 50 page publications under the public domain license.

50 Page Publications.

Book Title: Human Exposure and Environmental Impact

Book Author: E Braun; Richard D. Peacock; Glenn P. Forney; George W. Mulholland; Barbara C. Levin;

Book Abstract: Although these agents are typically employed in unoccupied sections of an aircraft, the possibility of human exposure still exists during handling, storage, and transport. Thus, it is important to know if the accidental release of the 12 agents in areas of typical occupancy would result in differing threats to life safety. At least two topics are important in assessing the impact of a potential release of an agent: 1) how does the agent distribute in an occupied space upon an accidental release, and 2) how does this release affect personnel who may be exposed? For the former, a series of tests was conducted to study the release of four of the twelve agents in a sealed compartment to measure the airborne concentration of agent that results from complete venting of containers of typical size into spaces of typical volume. These tests were augmented with field modeling to extend the range of the test results to other compartment geometries. For the latter, published toxicological results for chronic or acute exposure are summarized. It is important to note that in these tests, no humans were exposed.

Citation: NIST SP - 861

Keyword: halons; exposure; human beings; environmental effects; halon 1301; large scale fire tests; toxicity; accidents; compartments; experiments; temperature distribution; fire extinguishing agents; fire suppression; halon alternatives

See: NIST SP861, W. L. Grosshandler, et al., Editors

9. HUMAN EXPOSURE AND ENVIRONMENTAL IMPACT

Emil Braun, Richard D. Peacock, Glenn P. Forney, George W. Mulholland, and Barbara C. Levin
Building and Fire Research Laboratory

9.1 Potential Human Exposure

Although these agents are typically employed in unoccupied sections of an aircraft, the possibility of human exposure still exists during handling, storage, and transport. Thus, it is important to know if the accidental release of the 12 agents in areas of typical occupancy would result in differing threats to life safety. At least two topics are important in assessing the impact of a potential release of an agent: 1) how does the agent distribute in an occupied space upon an accidental release, and 2) how does this release affect personnel who may be exposed? For the former, a series of tests was conducted to study the release of four of the twelve agents in a sealed compartment to measure the airborne concentration of agent that results from complete venting of containers of typical size into spaces of typical volume. These tests were augmented with field modeling to extend the range of the test results to other compartment geometries. For the latter, published toxicological results for chronic or acute exposure are summarized. It is important to note that in these tests, no humans were exposed.

9.1.1 Full-Scale Tests. In order to evaluate the effects of an agent release in a closed compartment on potential human occupants, a series of experiments were conducted under full-scale geometries to determine

- agent distribution within the compartment, and
- temperature distribution within the compartment

for near (within the jet plume) and far field conditions over a short period of time, *i.e.*, acute exposure conditions. The test duration was 30 minutes. Temperature and gas concentrations at various levels within the compartment were measured during this "exposure" period. All agent release tests were conducted with the cylinder oriented such that the agent liquid was at the bottom of the cylinder and release occurred from the top. Cylinders were filled with a measured amount of agent and pressurized to a nominal 4,200 kPa with nitrogen gas. Since resources (i.e., time, funding, and agent availability) were limited for this phase of the project, tests were conducted on a subset of agents that spanned a broad range (greater than a factor of 3) of molecular weights and vapor pressures. While furnishing detailed data on the release process for a specific subset of agents, it was the also the intent of these experiments to provide data that could be used to verify the computational modeling of the release process. Therefore, in addition to halon 1301, only four agents were selected for evaluation. These agents are listed in Table 1 along with the amounts and initial pressurization states. Halon 1301 was used as a reference agent to determine if the observed conditions were significantly different than those currently experienced through accidental releases of halon 1301.

9.1.2 Instrumentation. A compartment, measuring 2.45 m by 3.66 m by 2.45 m high, was constructed and instrumented as shown in Figure 1. Because of an extended doorway entrance the

Table 1. Agents tested in the full-scale geometry

Compound	Formula	Molecular Weight	Vapor Press. @ 25 °C (MPa)	Test Mass (g)
FC-116	C_2F_6	138	--[a]	465
Halon 1301	CF_3Br	149	1.61	420
				391
HCFC-22	CHF_2Cl	87	1.05	189
				297
HFC-227	C_3HF_7	170	0.47	108
HFC-32/HFC-125	CH_2F_2/C_2HF_5	67	1.67	227

[a]above critical pressure

total volume of the compartment was 22.7 m³. A 1 liter stainless steel flask identical to those used in the stability tests was filled with test agent at approximately 4,200 kPa. The flask was attached to a support stand located in the center of the compartment. The support stand consisted of a swagelock mating nut, pneumatic actuator and valve, and a 10 mm ID tube that extended to a release point 0.9 m from the floor (Figure 2). Each experiment involved installing the charged agent cylinder into the support frame, sealing the compartment, and actuating the pneumatic valve to release agent into the compartment. The agent cylinder was left open through the duration of the test, 30 minutes. Thermocouple data were collected every ten seconds and evacuated flasks were filled in a specified pattern (Table 2) to ensure that if stratification or settling occurred it could be detected.

Three types of measurements were made during the release of each agent. Thermocouple measurements were made to determine the thermal effects of a rapid release of agent. Gas concentrations were measured using evacuated flasks as well as FTIR measurements to determine agent distribution in the compartment.

9.1.2.1 Temperature. Thermocouples were located on two vertical strings. One string of thermocouples was located in a corner of the compartment, 0.6 m from each adjoining wall. A second string of thermocouples was located 0.5 m north of the release point. Each thermocouple string consisted of eight thermocouples spaced 0.3 m apart starting at the ceiling. Thermocouples were also located 0.3, 0.6, and 0.9 m above the release point as well as one thermocouple 0.15 m above and 0.08 m north and another thermocouple 0.3 m above and 0.15 m north of the release point. Thermocouple data was collected every 10 seconds.

9.1.2.2 Gas Concentration. Using evacuated flasks, gas were sampled at three locations in the compartment. One sampling probe was placed 1.23 m above the release point (0.31 m from the ceiling). The other two sampling probes were located 0.46 m from the south wall. For tests involving agents FC-116, halon 1301, and HCFC-22 the two probes were located 0.92 m and 1.98 m, respectively, from the ceiling. For agents halon 1301, HCFC-22, HFC-227, and HFC-32/HFC-125 the lower probe was placed 44.5 mm from the floor (approximately 2.41 m from the ceiling). This was done to determine if agent settling was occurring. Each sampling probe was connected to a four

SECTION 9. HUMAN EXPOSURE AND ENVIRONMENTAL IMPACT

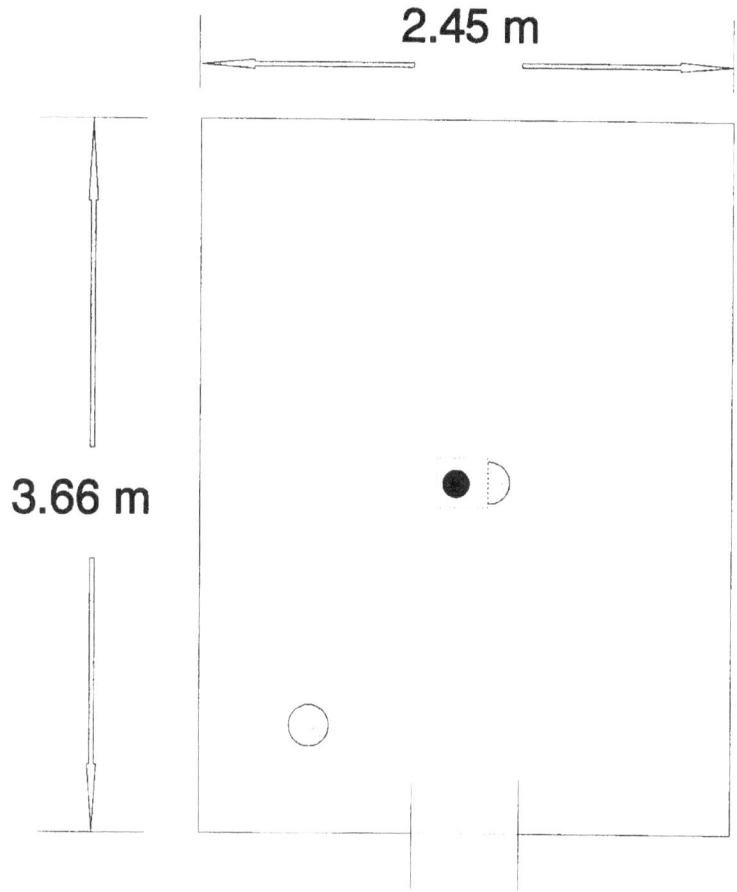

Figure 1. Floor plan and instrument layout for halon 1301 replacement release tests.

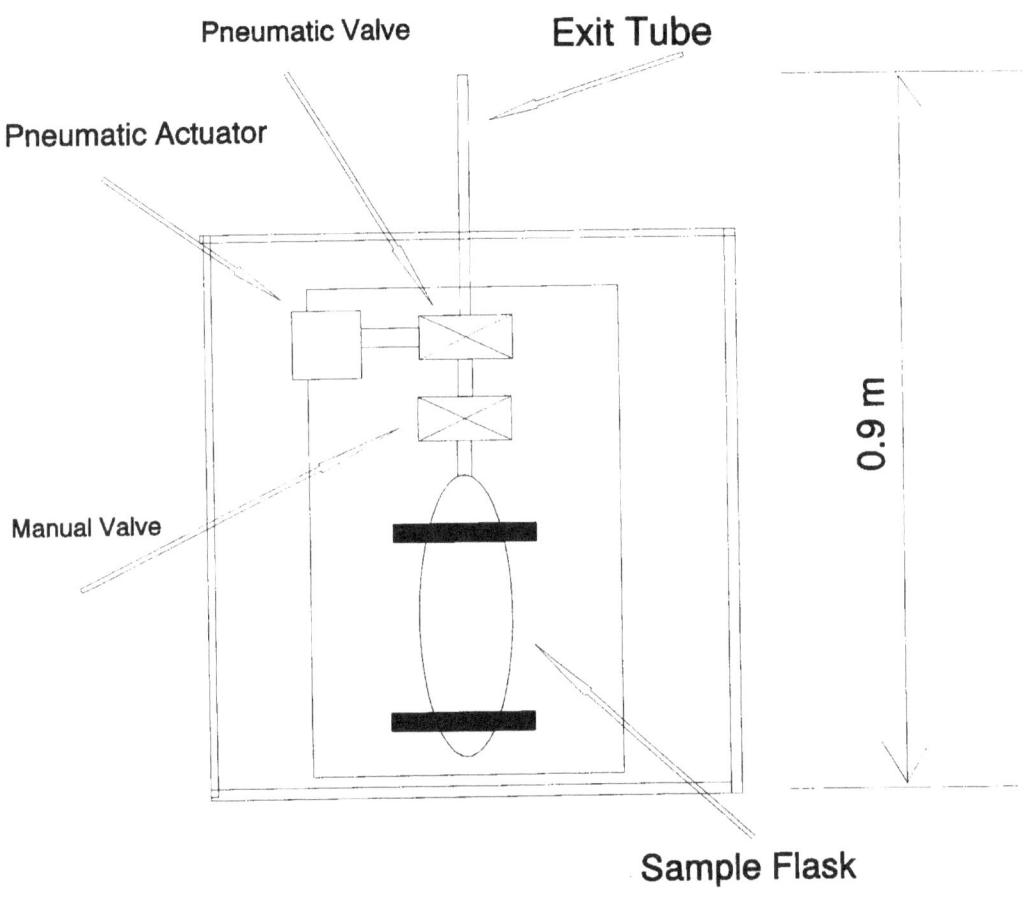

Figure 2. Schematic of agent sample flask, support frame, manual valve, pneumatic valve, and exit tube in relationship to the floor of the compartment.

SECTION 9. HUMAN EXPOSURE AND ENVIRONMENTAL IMPACT

flask manifold. Room atmosphere was drawn through the manifold by a vacuum pump operating at 2.5 l/m. In order to reduce atmospheric mixing in the compartment, gas samples drawn through each manifold were returned to the compartment at the same level from which they were taken at approximately the same location. Prior to a test, 12 flasks, 125 ml each, were evacuated. Four flasks were connected to each manifold. The vacuum pump for each manifold operated continuously, except during the time it took to fill a flask. The content of each flask was analyzed by a Hewlett Packard HP5730A gas chromatograph equipped with a flame ionization detector (FID). A column packed with 60/80 Carbopack B/ 5% Fluorcol was used to separate the agent from the background atmosphere. For the two component agent (HFC-32/HFC-125), peaks were detected representing each component. The carrier gas was nitrogen flowing at 30 ml/min. Each injection represented 500 μl of gas drawn from a flask. An approximate calibration was determined for each agent by injecting 500 μl samples drawn from a flask containing a known amount of agent dispersed in a know volume of air. Measurement errors associated with this calibration procedure consist of the error in measuring the volume of the calibration flask, the weight of the calibration flask before and after the addition of the agent, and the volume of gas drawn from the flask for injection into the GC carrier gas. A balance with a reading resolution of 0.1 mg was used to determine the mass of agent added to the flask. The balance was also used to determine the volume of the flask by measuring the mass of water needed to completely fill the flask. An error of 0.1 mg in the mass of agent in 125 ml flask represents a variation of approximately 0.8 mg/l in agent concentration or, for agent concentration range of 5 to 21 mg/l, represents a variation of about 4 to 15%. Errors associated with the GC system are a combination of syringe volume variation and GC system variation. These errors can be approximated by computing the variation in integrator area calculations. This error was found to vary from 1 to 4%. Combining these errors results in a final error approximation ranging from 5 to 19%.

Relative gas concentrations were also monitored in-situ with a Fourier Transform InfraRed (FTIR) spectrometer. The FTIR was relocated for each test depending on the intent of the measurement. For tests of FC-116, halon 1301 (test 1301-1), and HCFC-22 (test 22-1), the FTIR was located such that the sampling beam was approximately 1.2 m from the ceiling aimed diagonally across the compartment. For tests of HFC-32/HFC-125 and HCFC-22 (test 22-2), the FTIR was located 0.45 m from the ceiling with the beam parallel to and 0.52 m from the north wall. For the retest of halon 1301 (test 1301-2), the FTIR was placed along the north wall 0.05 m from the floor. While the FTIR provided better time resolution of agent concentration than the evacuated flasks, spatial resolution was provided by the data obtained from the use of evacuated flasks to sample the enclosure atmosphere as specified above. It was hoped that comparable trends would be observed by both measurement techniques.

9.1.3 Experimental Results

9.1.3.1 Temperature. The data from the thermocouple string located in the corner of the compartment displayed only small deviations from ambient conditions, less than 1 °C. These far field temperature measurements indicate that the rapid release of any of the agents at the loading concentrations used in these experiments produced little deviation from ambient conditions. Near field measurements below the point of release also showed deviations of less than 1 °C from ambient conditions for all tested agents. Data from these locations will not be shown or discussed further. Figures 3 to 9 show the temperature data for each test at the following three general sets of locations:

Table 2. Description of evacuated flask location, sampling sequence and timing

Sequence Number	Location Distance from Ceiling (m)		Designation	Sample Time (s)
1	0.31[a]		I.1	30
2	0.31[a]		I.2	90
3	0.91[b]		II.1	120
4	1.98[b,c]	2.41[b,d]	III.1	180
5	0.31[a]		I.3	300
6	0.91[b]		II.2	420
7	1.98[b,c]	2.41[b,d]	III.2	480
8	0.91[b]		II.3	600
9	1.98[b,c]	2.41[b,d]	III.3	720
10	0.31[a]		I.4	1320
11	1.98[b,c]	2.41[b,d]	III.4	1500
12	0.91[b]		II.4	1680

[a] Centered over exit tube
[b] 0.46 m from South Wall and 1.43 m from East Wall
[c] Height for agent release test: halon 1301, FC-116, HCFC-22
[d] Height for agent release test: halon 1301, HCFC-22, HFC-227, HFC-32/HFC-125

1. directly above the exit tube,
2. for a thermocouple 0.15 m above and 0.08 m north and another thermocouple 0.3 m above and 0.15 m north of the release point, and
3. for the upper four thermocouples 0.5 m north of the release point.

To provide the most visually clear set of data for each test, we have used different temperature ranges in the graphs. Care must therefore be exercised when comparing results between agents.

Figure 10 schematically shows the relationship between the thermocouples in the vicinity of the exit tube, including all of the thermocouples displayed in Figures 3 to 9. For all tests, the largest deviations from ambient occurred directly above the exit tube with the minimum temperature recorded by the thermocouple closest to the exit tube. Based on the thermocouple data in Figures 3 to 9, the release pattern can clearly be discerned. In every "A" graph of Figures 3 to 9, the change in minimum temperature decreases as the distance from the exit tube increases. In the "C" graphs, the deviation increases as the height above the exit tube level increases. Furthermore, the data from the two off-axis thermocouples displayed in the graphs labeled "B" show no significant response to the

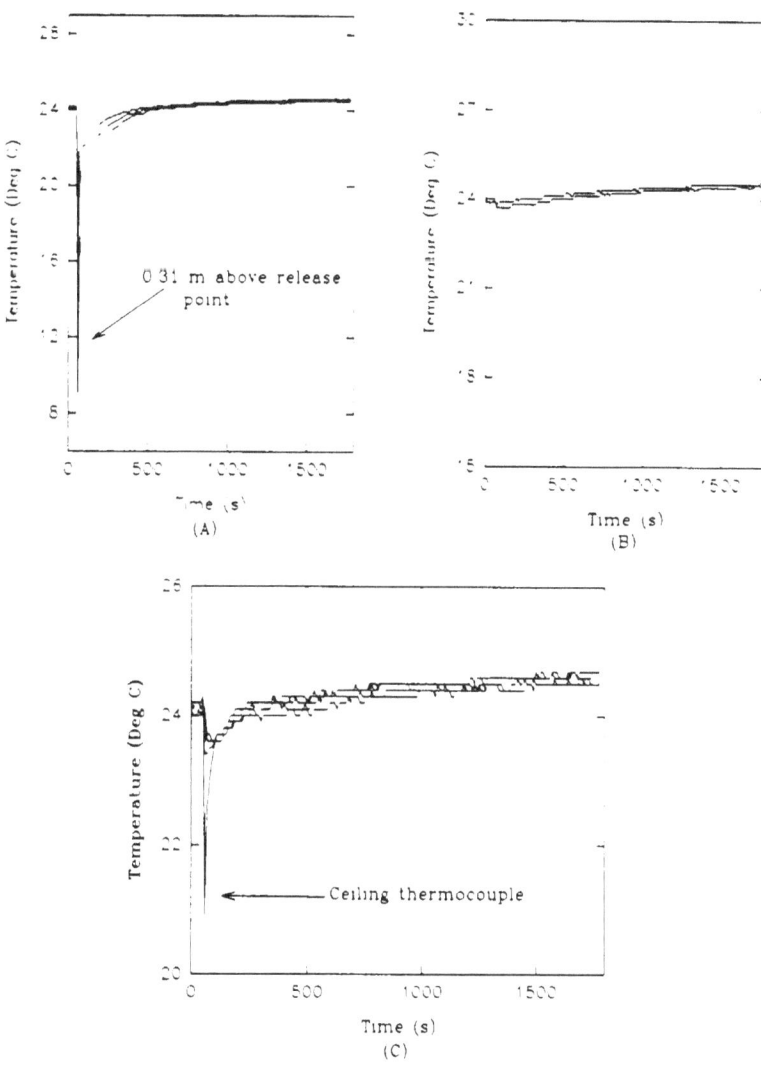

Figure 3. Temperature data FC-116: A-at 0.3, 0.6, and 0.9 m above tube; B-one 0.15 m above, 0.08 m offset and one 0.3 m above, 0.15 m offset; C-four 0.5 m offset, 0.0, 0.31, 0.62, and 0.93 m from ceiling.

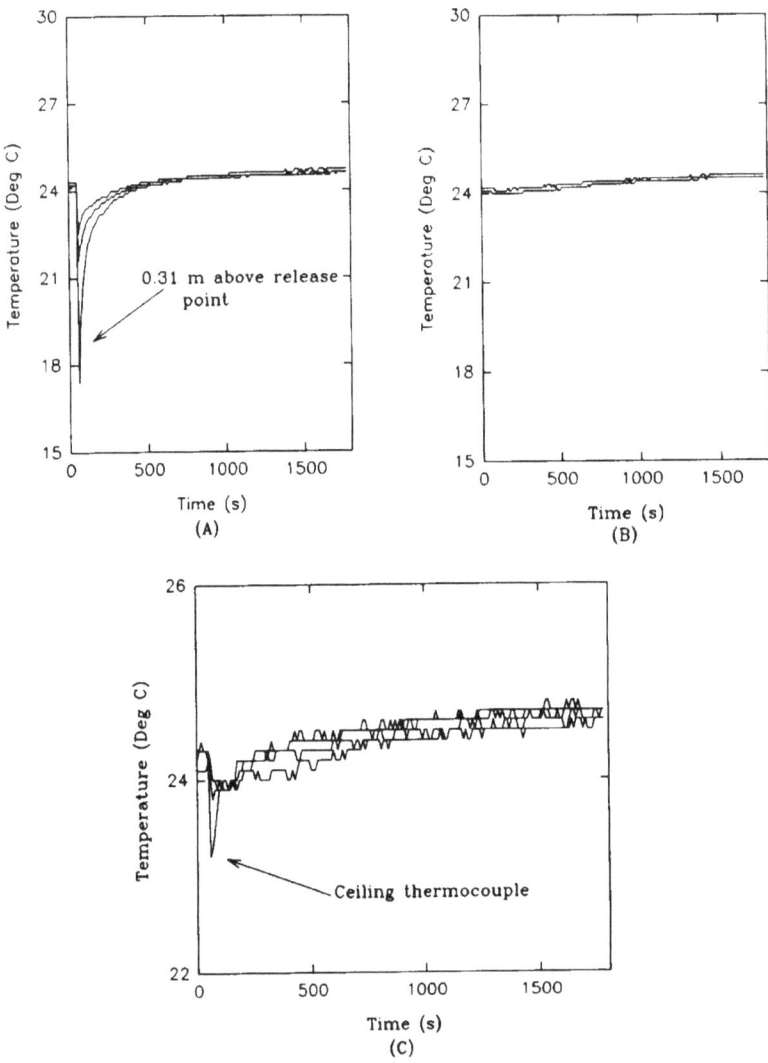

Figure 4. Temperature data Halon 1301, test 1: A-at 0.3, 0.6, and 0.9 m above tube; B-one 0.15 m above, 0.08 m offset and one 0.3 m above, 0.15 m offset; C-four 0.5 m offset, 0.0, 0.31, 0.62, and 0.93 m from ceiling.

SECTION 9. HUMAN EXPOSURE AND ENVIRONMENTAL IMPACT

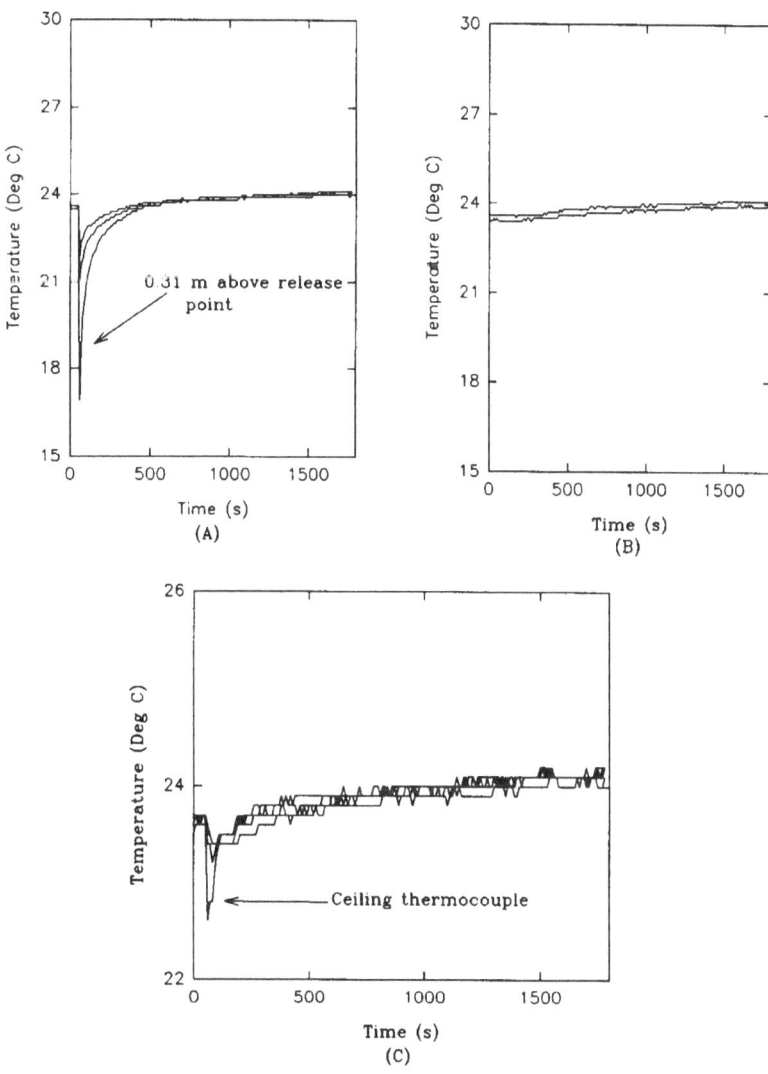

Figure 5. Temperature data Halon 1301, test 2: A-at 0.3, 0.6, and 0.9 m above tube; B-one 0.15 m above, 0.08 m offset and one 0.3 m above, 0.15 m offset; C-four 0.5 m offset, 0.0, 0.31, 0.62, and 0.93 m from ceiling.

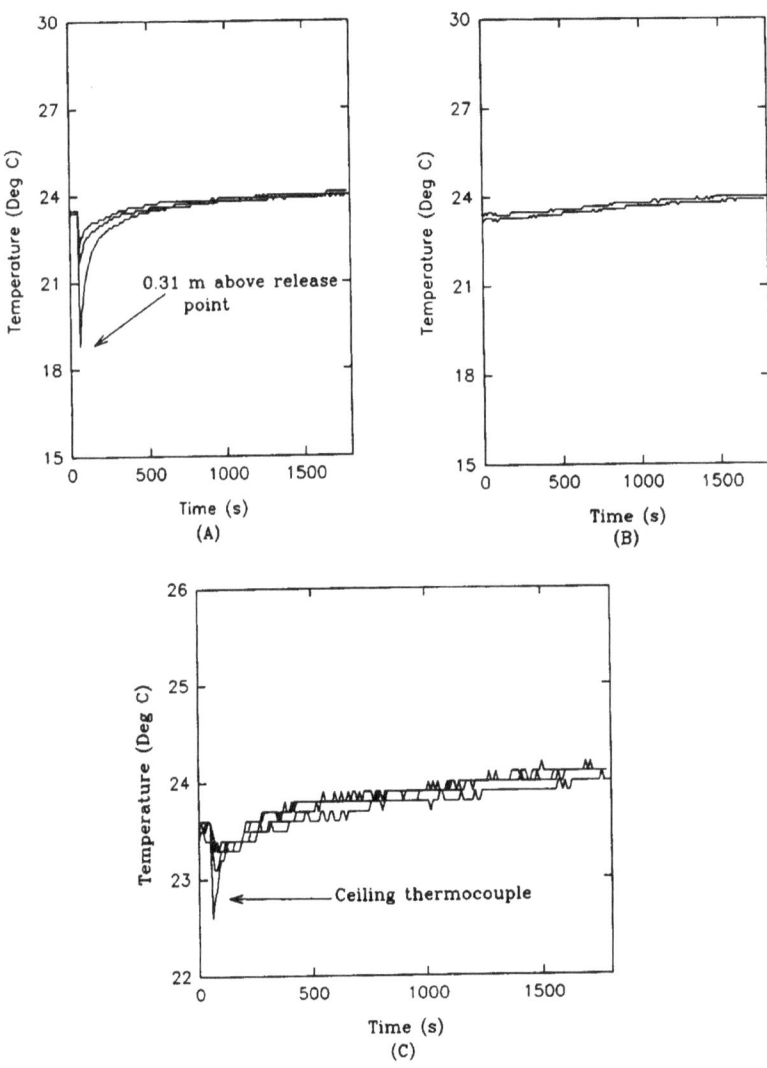

Figure 6. Temperature data HCFC-22, test 1: A-at 0.3, 0.6, and 0.9 m above tube; B-one 0.15 m above, 0.08 m offset and one 0.3 m above, 0.15 m offset; C-four 0.5 m offset, 0.0, 0.31, 0.62, and 0.93 m from ceiling.

SECTION 9. HUMAN EXPOSURE AND ENVIRONMENTAL IMPACT

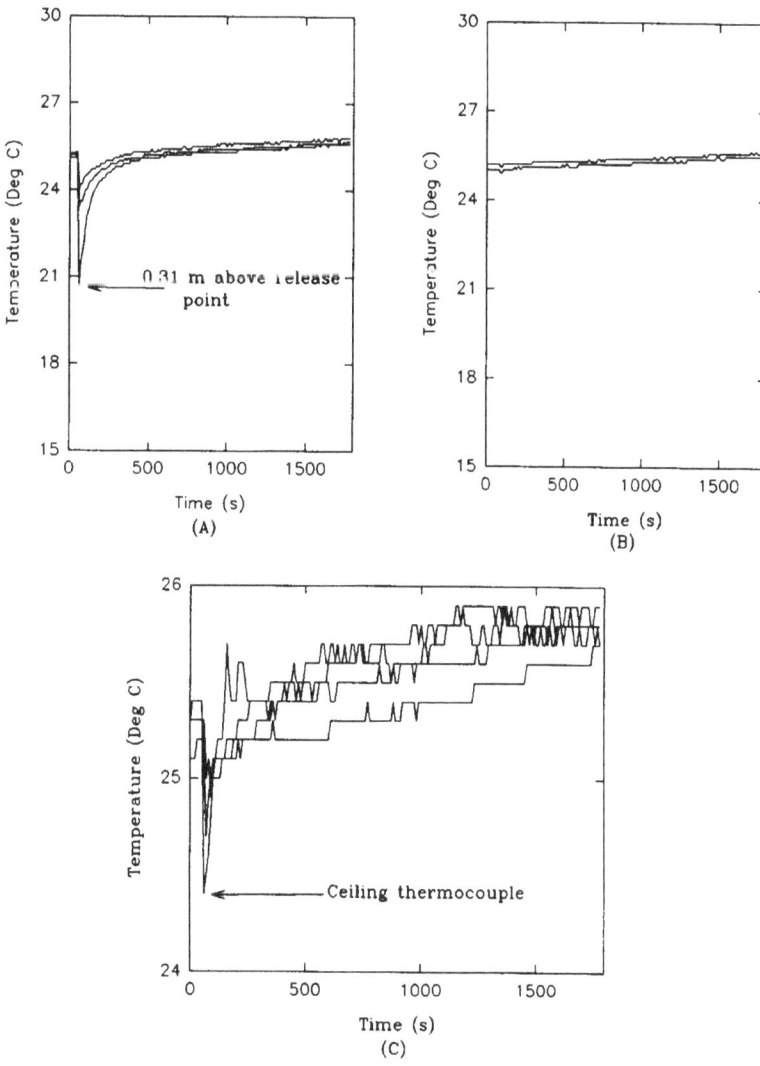

Figure 7. Temperature data HCFC-22, test 2: A-at 0.3, 0.6, and 0.9 m above tube; B-one 0.15 m above, 0.08 m offset and one 0.3 m above, 0.15 m offset; C-four 0.5 m offset, 0.0, 0.31, 0.62, and 0.93 m from ceiling.

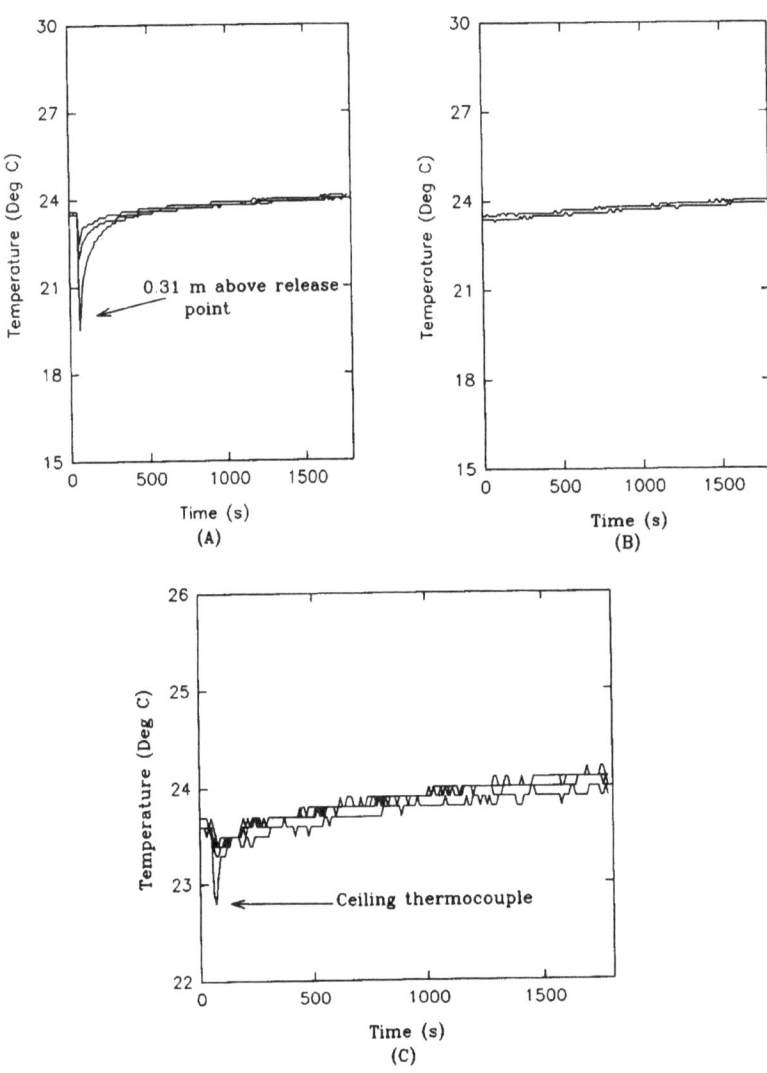

Figure 8. Temperature data HFC-227: A-at 0.3, 0.6, and 0.9 m above tube; B-one 0.15 m above, 0.08 m offset and one 0.3 m above, 0.15 m offset; C-four 0.5 m offset, 0.0, 0.31, 0.62, and 0.93 m from ceiling.

SECTION 9. HUMAN EXPOSURE AND ENVIRONMENTAL IMPACT

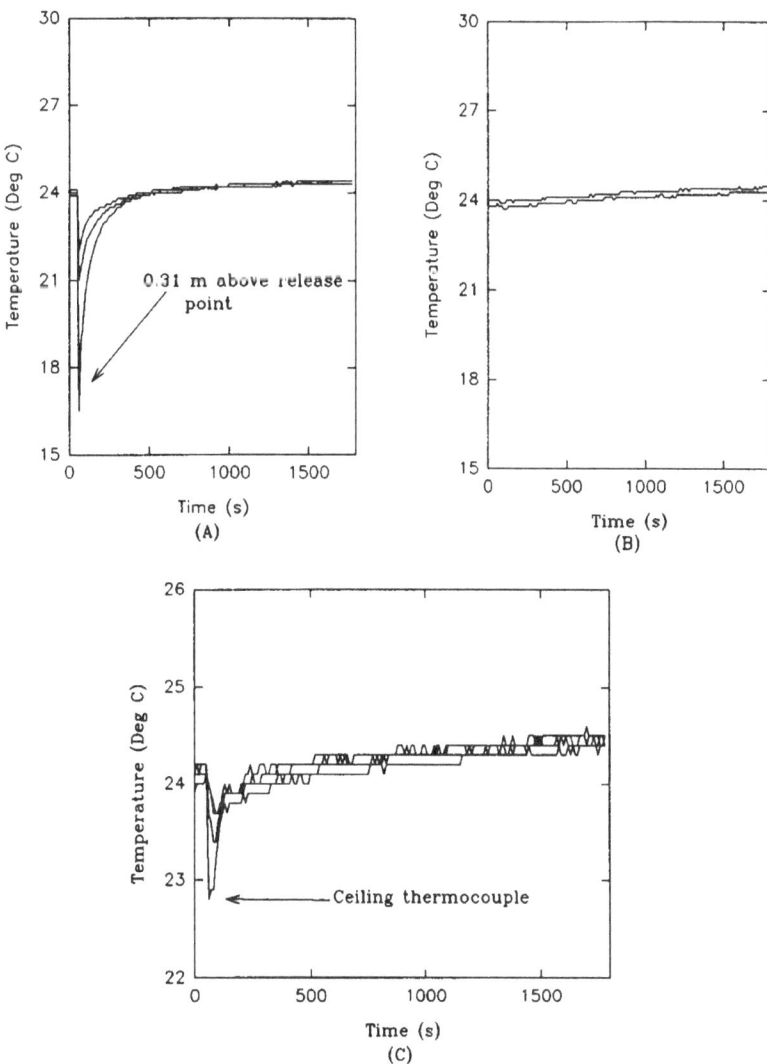

Figure 9. Temperature data HFC-32/125: A-at 0.3, 0.6, and 0.9 m above tube; B-one 0.15 m above, 0.08 m offset and one 0.3 m above, 0.15 m offset; C-four 0.5 m offset, 0.0, 0.31, 0.62, and 0.93 m from ceiling.

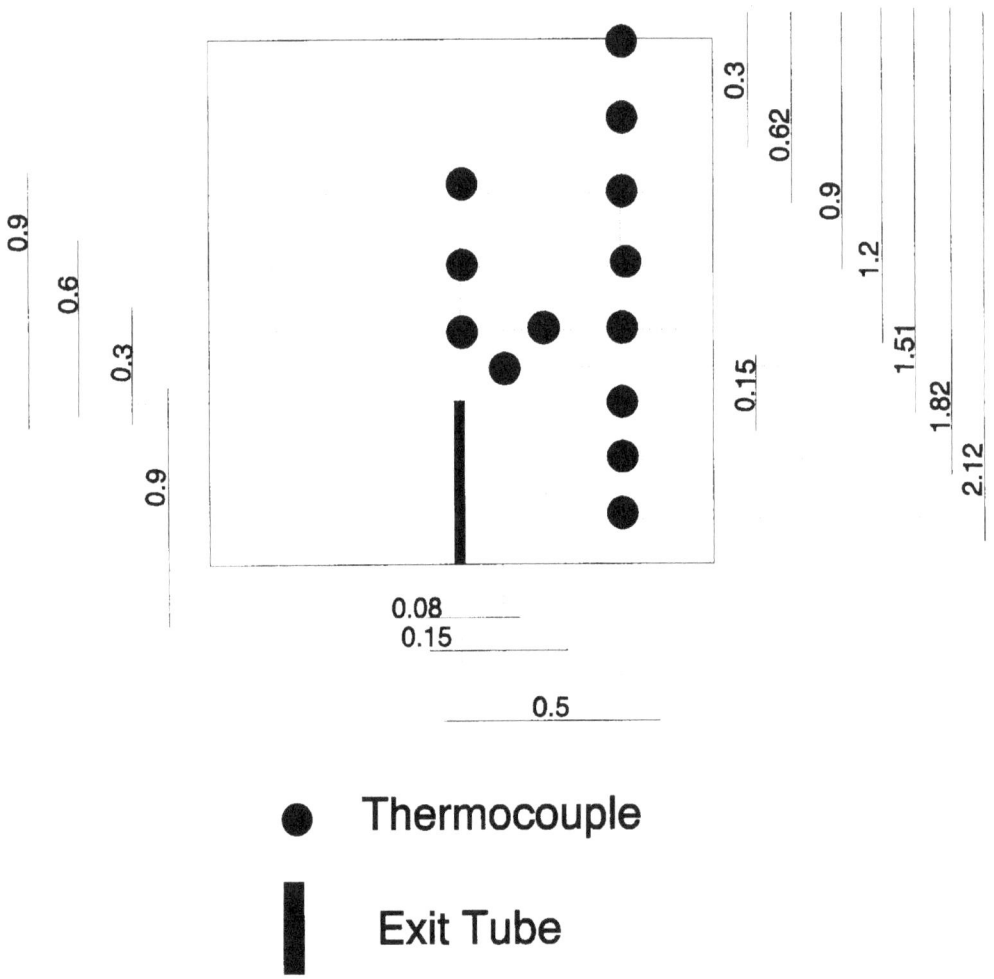

Figure 10. Schematic drawing showing the relationship of the near field thermocouples to the exit tube in the north-south vertical plane of the agent sample flask.

SECTION 9. HUMAN EXPOSURE AND ENVIRONMENTAL IMPACT

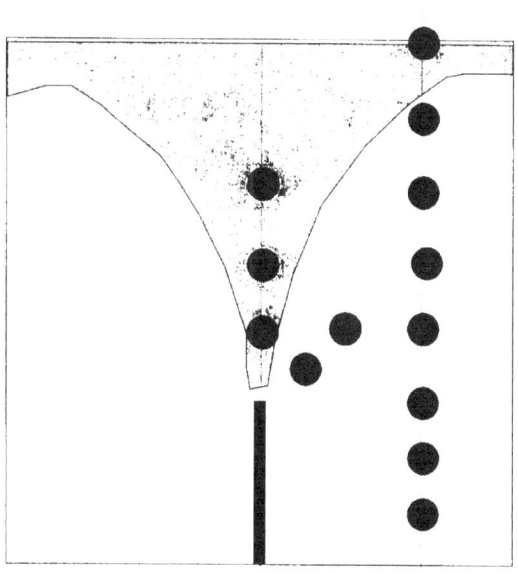

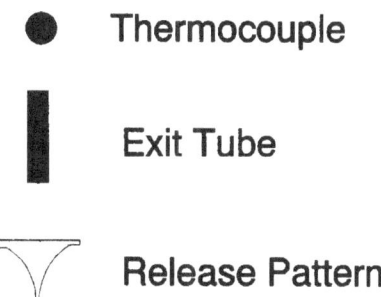

Figure 11. Theoretical representation of agent plume in relationship to the thermocouples located in the north-south vertical plane of the agent sample flask.

Table 3. Summary of the extent of temperature depression of two thermocouple locations

Compound		Thermocouple @ 0.31 m above exit tube		Thermocouple @ ceiling 0.5 m north of exit tube	
		Maximum Depression (°C)	Time to Recover to 90% of Ambient (s)	Maximum Depression (°C)	Time to Recover to 90% of Ambient (s)
FC-116		15.8	170	3.3	130
Halon 1301	1	6.8	140	1.1	390
	2	6.6	180	1.1	130
HCFC-22	1	4.6	190	1.0	170
	2	4.4	135	1.0	185
HFC-227		4.0	180	0.8	80
HFC-32/HFC-125		7.4	185	1.4	90

sudden release of agent. This suggests that the release plume is very narrow and quickly reached the compartment ceiling. The plume spreads rapidly along the ceiling and expands down towards the floor of the compartment. This is shown idealized in Figure 11. Based on the thermocouple data, this process occurs rapidly.

Table 3 summarizes the magnitude of the maximum depression from ambient temperature for the thermocouple located 0.3 m above the exit tube and the time to return to within 90% of ambient conditions. The same data are also tabulated for the ceiling thermocouple of the 0.5 m north thermocouple string. Maximum depression directly above the exit tube ranged from 15.8 °C for agent FC-116 to 4.0 °C for agent HFC-227. The maximum depression of the ceiling thermocouple was 3.3 °C for FC-116 to 0.8 °C for HFC-227. The 90% recovery time to near ambient conditions, with one exception, ranges from 80 s to 190 s. The recovery time data displays a classically exponential increase in temperature consistent with the natural convective heating of an object.

9.1.3.2 Concentration. In-situ FTIR and atmospheric samples captured in evacuated flasks represented the two methods employed in this study to determine agent concentration in the compartment. The FTIR apparatus is a line-of-sight measurement of the gas concentration of the released agents. The evacuated flasks are point measurements made at specific time intervals and, because three distinct sampling probes were employed, specific locations as previously described.

Figures 12 to 18 show the measured gas concentrations as a function of time for the three probe locations. Also noted on each figure (dashed line) is the average agent concentration in the compartment assuming (a) complete release of the agent from the cylinder, (b) even distribution within the compartment, and (c) no losses from the compartment.

While an effort was made to ensure that the compartment was completely sealed, the rapid decay in the early data (first half of the test period) indicates that significant leakage may have occurred.

SECTION 9. HUMAN EXPOSURE AND ENVIRONMENTAL IMPACT

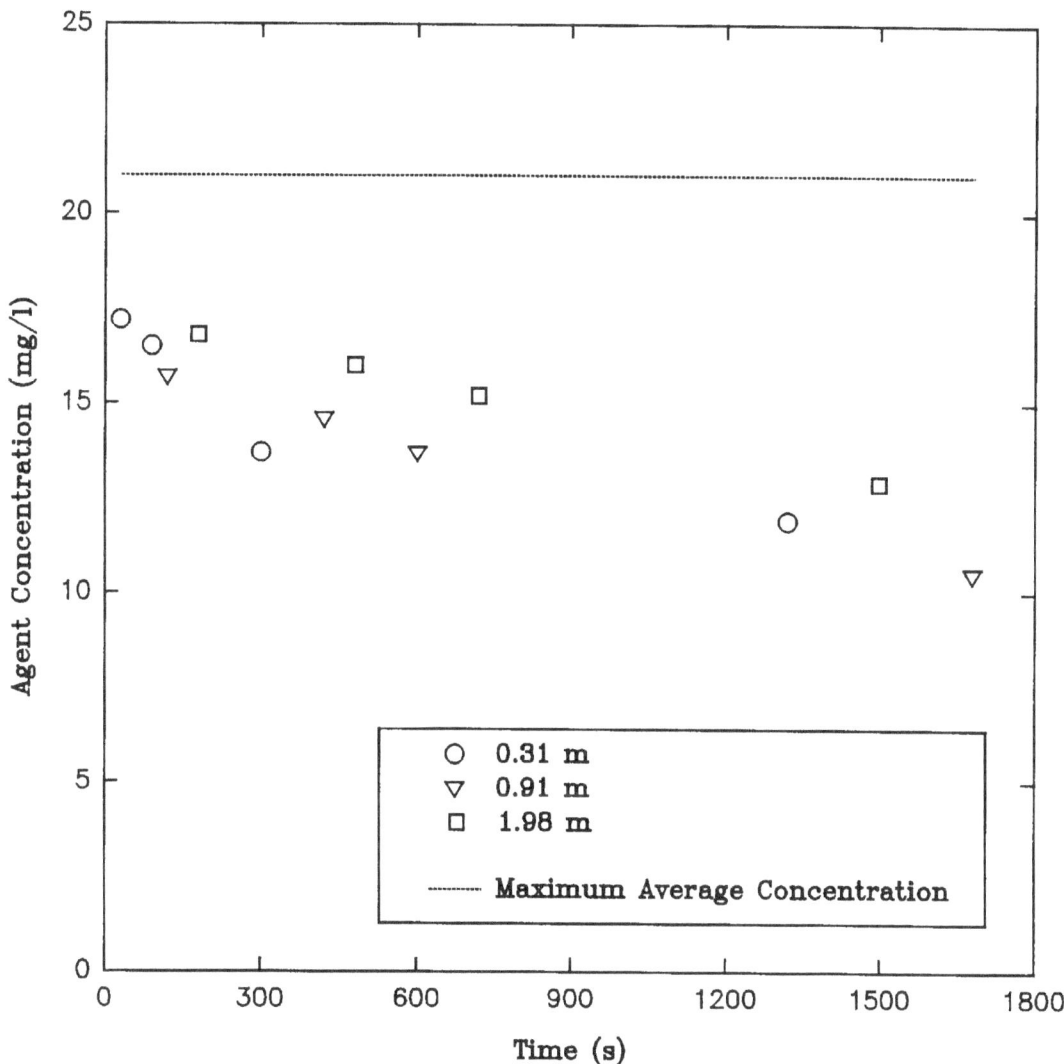

Figure 12. Agent concentration for agent FC-116 as a function of position and time with the maximum average concentration of agent shown (----) assuming uniform distribution within the compartment.

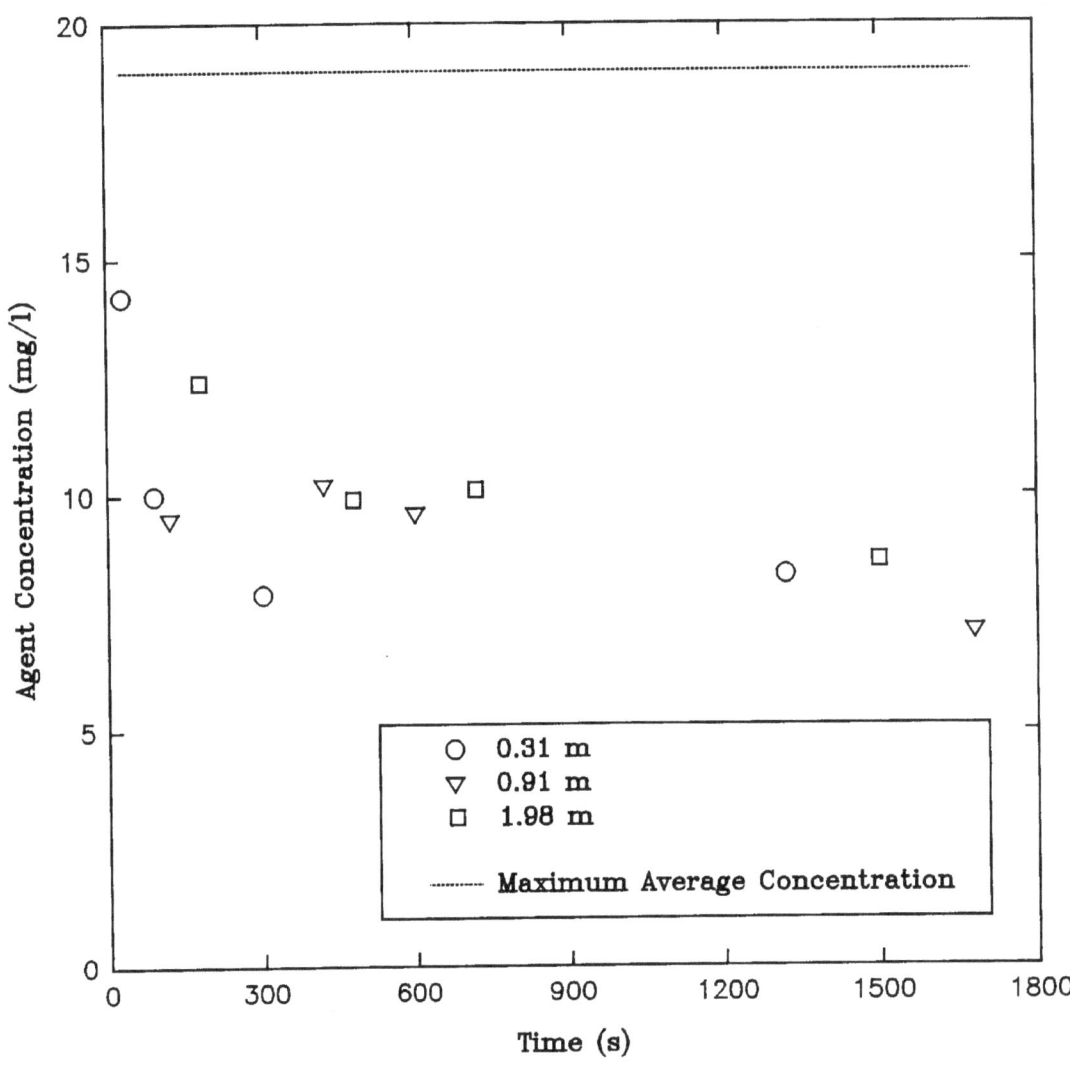

Figure 13. Agent concentration for halon 1301 test 1 as function of position and time with the maximum average concentration of agent shown (----) assuming uniform distribution within the compartment.

SECTION 9. HUMAN EXPOSURE AND ENVIRONMENTAL IMPACT

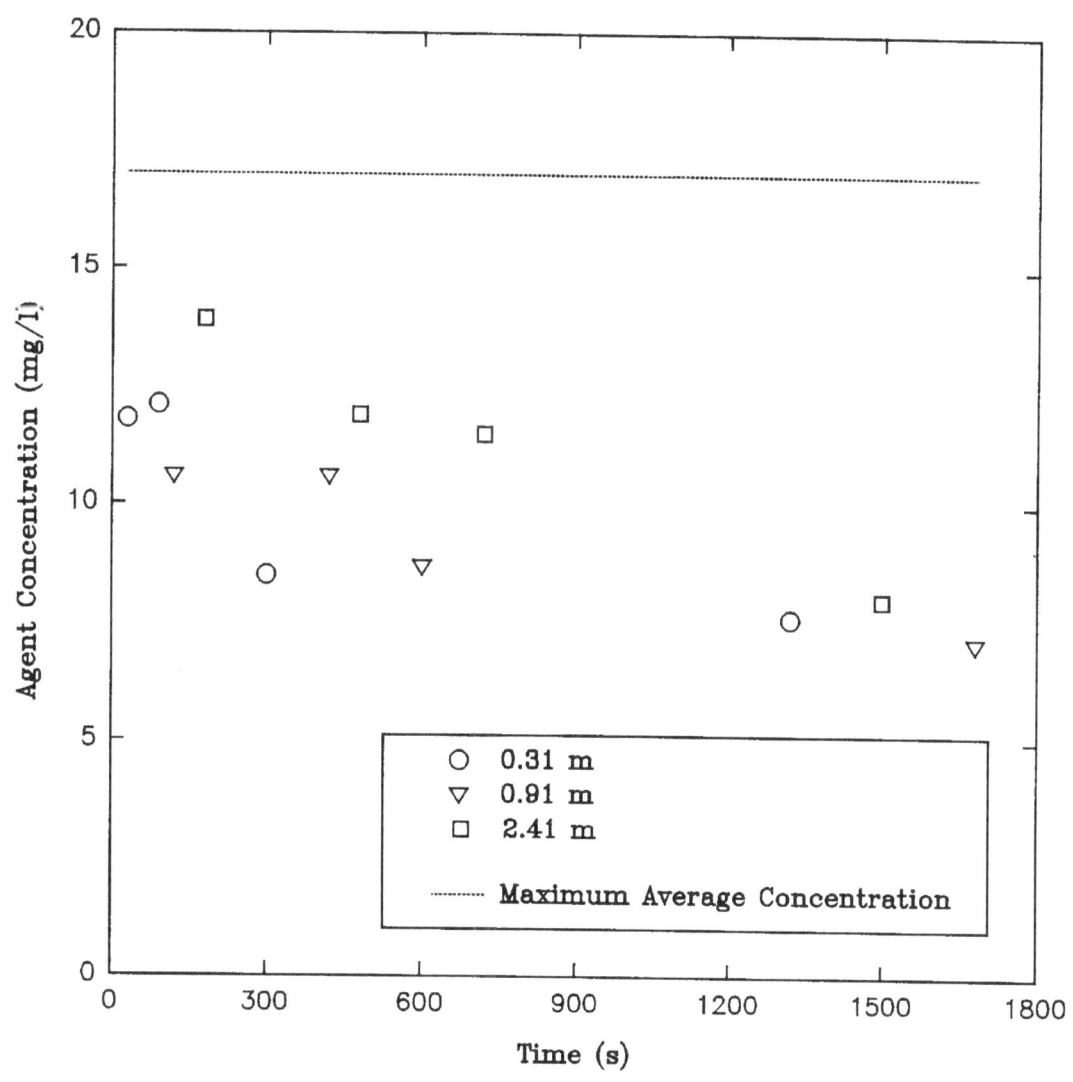

Figure 14. Agent concentration for halon 1301 test 2 as a function of position and time with the maximum average concentration of agent shown (----) assuming uniform distribution within the compartment.

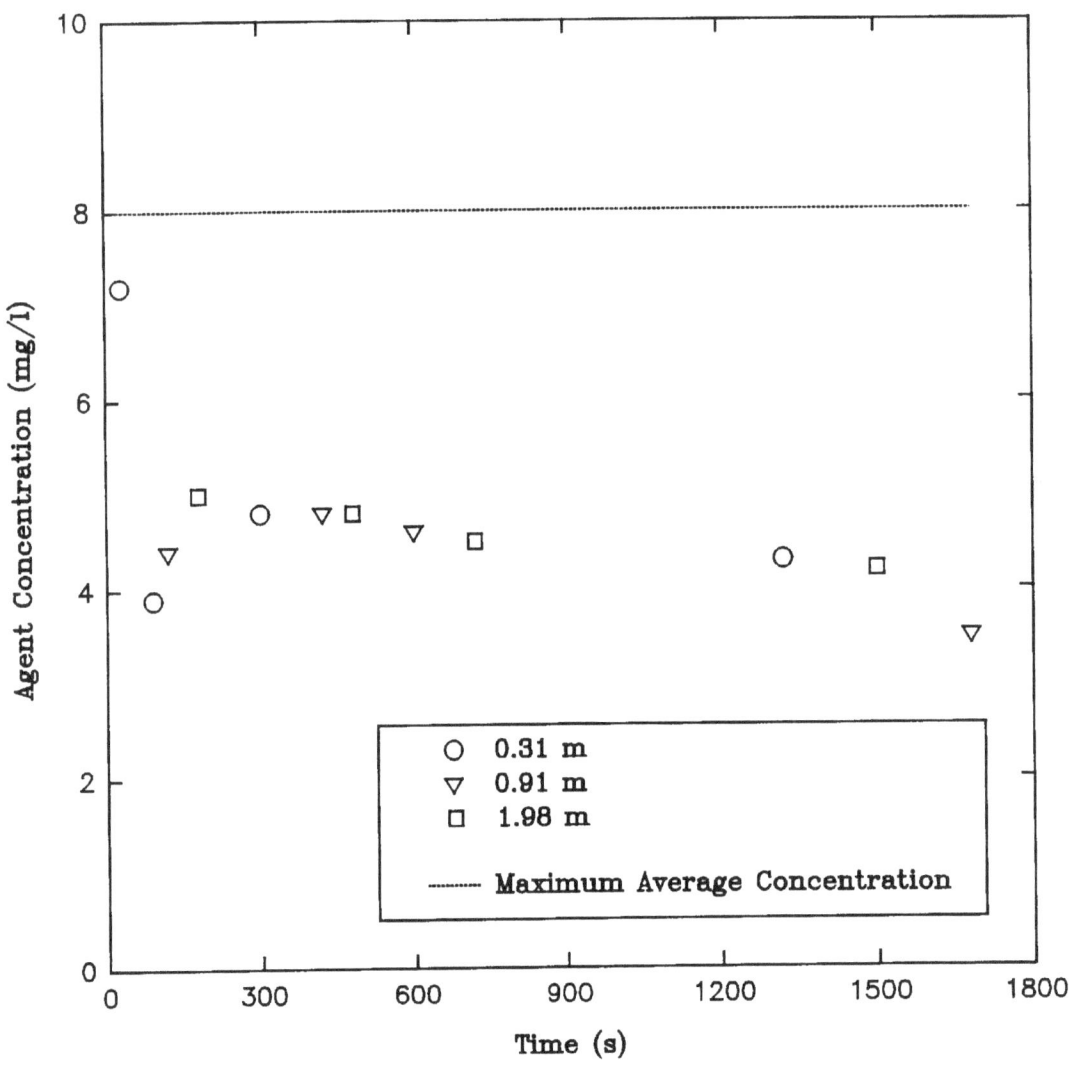

Figure 15. Agent concentration for HCFC-22 test 1 as a function of position and time with the maximum average concentration of agent shown (----) assuming uniform distribution within the compartment.

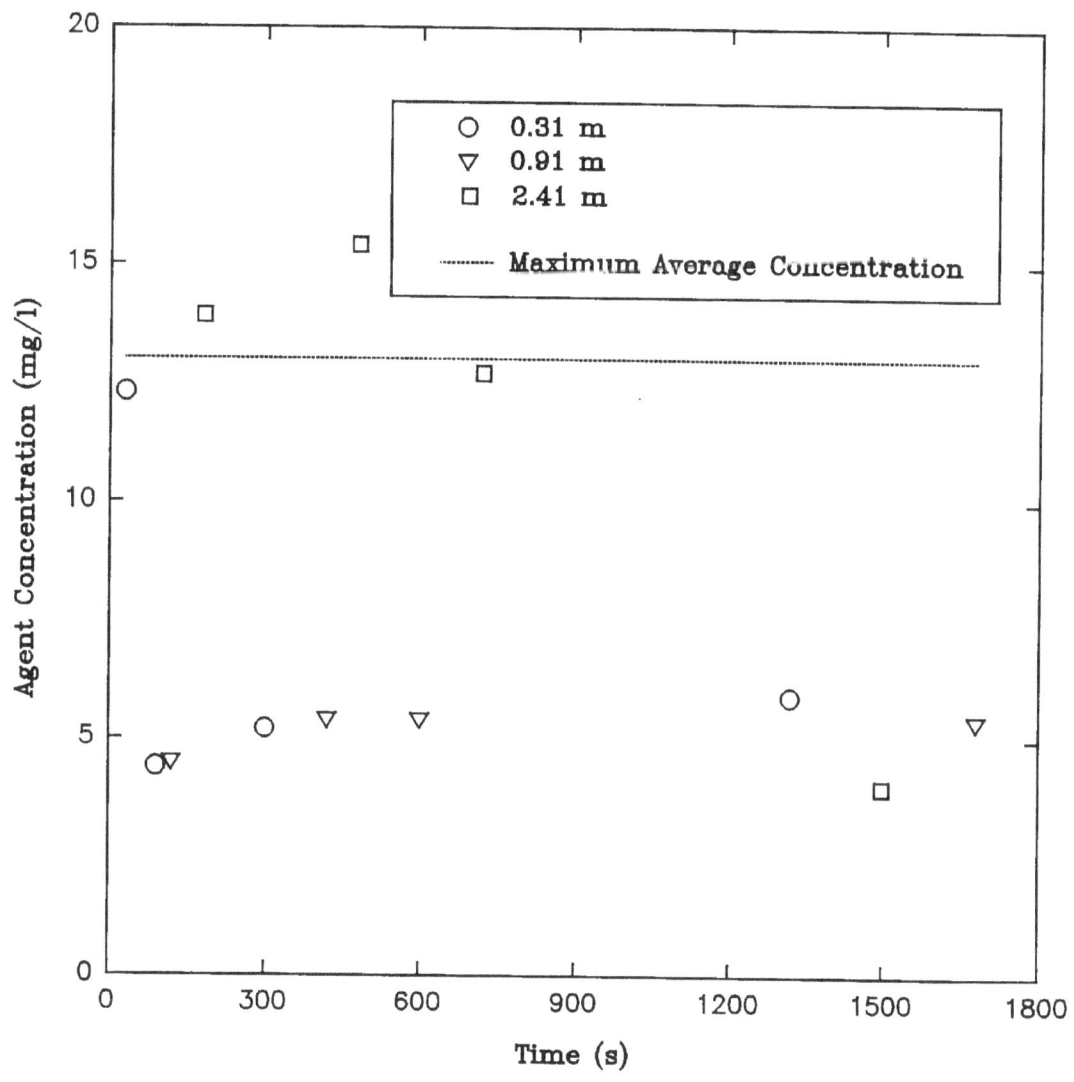

Figure 16. Agent concentration for HCFC-22 test 2 as a function of position and time with the maximum average concentration of agent shown (----) assuming uniform distribution within the compartment.

786 SECTION 9. HUMAN EXPOSURE AND ENVIRONMENTAL IMPACT

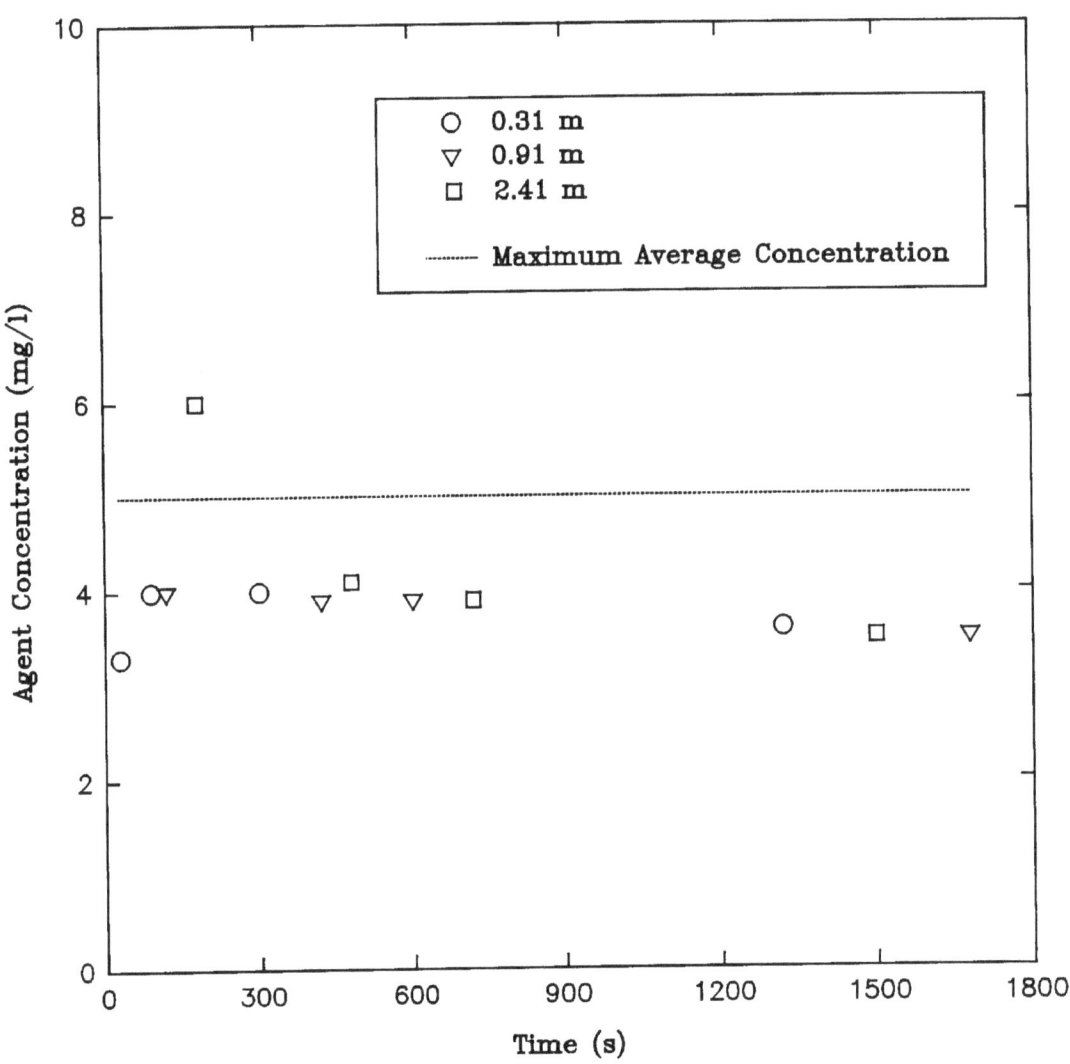

Figure 17. Agent concentration for agent HFC-227 as a function of position and time with the maximum average concentration of agent shown (----) assuming uniform distribution within the compartment.

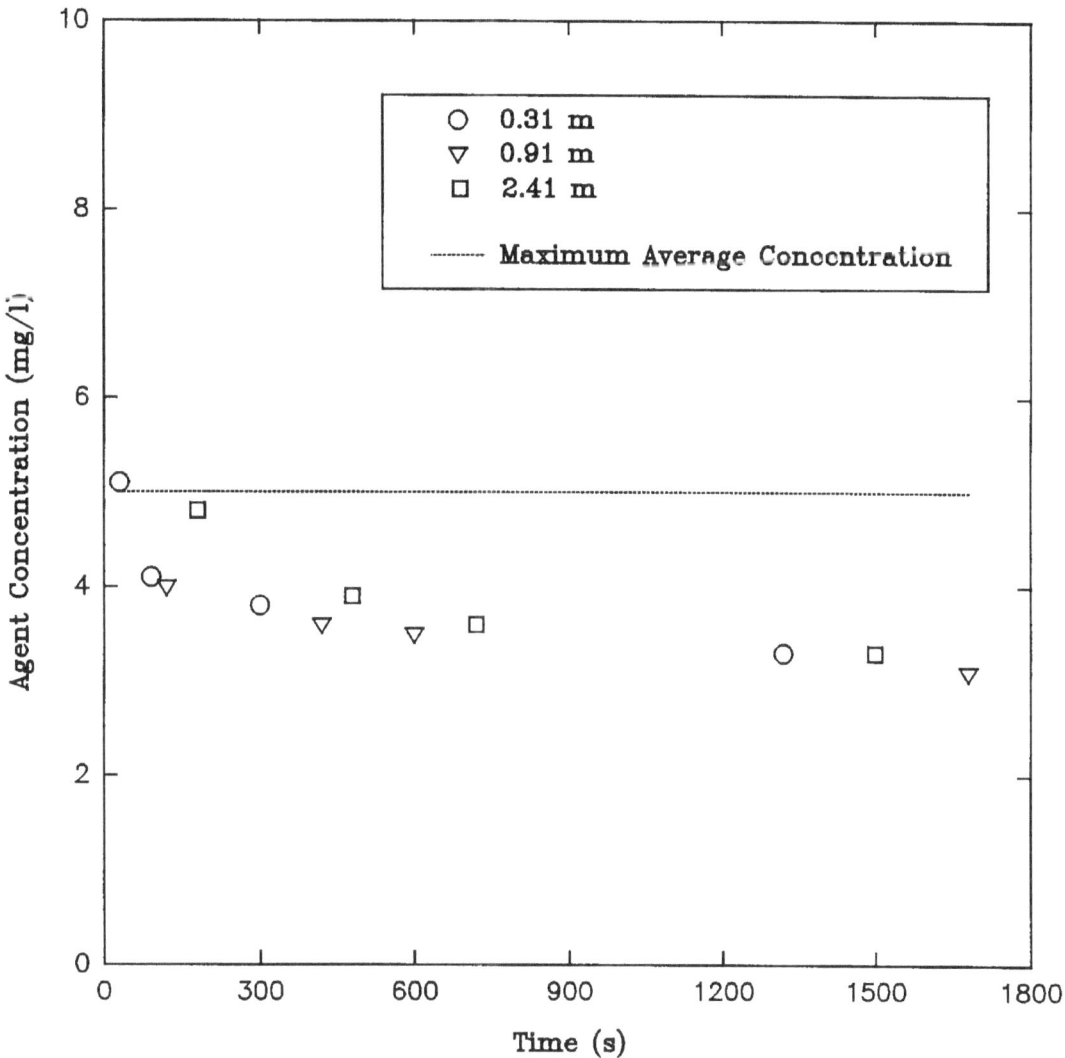

Figure 18. Agent concentration for agent HFC-32/125 as a function of position and time with the maximum average concentration of agent shown (----) assuming uniform distribution within the compartment.

However, once the turbulent action of the initial agent release had dissipated, further leakage would only affect the final compartment concentration and not the ultimate distribution of agent within the compartment.

Table 4 summarizes the average agent concentration for each release test, based on the average of the last sampling time period for the three locations (this represents a time period of 6 minutes) as well as the theoretical average concentration. The coefficient of variation, computed as the ratio of the standard deviation to the sample mean, for all tests was 3 to 19%. This was within the measurement variation of the GC analysis procedure. The data also shows that the final measured average agent concentration was always less than the theoretical average value. This could be due to (a) not all of the agent being released into the compartment (*i.e.*, residual agent remaining in cylinder after the discharge) or (b) leakage of agent from the compartment.

In general, the data show that the initial concentration in the compartment was below the expected average agent concentration assuming complete release of agent and that the concentration at a given level within the compartment decreased as a function of time. All the data indicate that an average steady-state concentration was reached within the 30 minute test period.   Table 5 compares the final average agent concentration within the compartment with the 300 s agent concentration 0.31 m from the ceiling. The data show that the 300 s data was never more than 20.0% from the final compartment average agent concentration. In 4 out of 7 cases the 300 s data was less than 15% above the final average concentration. Within measurement errors, the data appear to indicate that a uniform agent distribution was developed within a short period of time (approximately 300 s).

In the 30 min duration of the experiments, it is not likely that significant amounts of agent would be left in the cylinders. Leakage from the compartment is likely, especially during the initial, high velocity release of the agent. The tests, however, indicate that computing the theoretical average concentration would represent a conservative estimate of occupant exposure conditions shortly after an accidental release of agent.

A comparison of the data for agent HCFC-22, Figures 15 and 16, shows how quickly agent concentrations can change at various levels in the test compartment. While the data for the two release tests show a dramatic difference for the first 600 s after agent release, the final measured average agent concentration did not significantly differ. This was probably due to a combination of convective mixing, diffusion, and compartment leakage.

Tests on halon 1301 (Figures 13 and 14) were also repeated to determine the effects of probe location on the measured agent concentrations. For the data in Figure 13, the lower probe was 1.98 m from the ceiling. For Figure 14, the lower probe was located 2.41 m from the ceiling (nearly at the floor). Unlike HCFC-22 agent release tests, which showed a short term stratification of agent, the halon 1301 tests showed no such stratification. There was no significant difference between the data at all levels between these two tests.

To further determine the time response of agent mixing and, therefore, concentration within the compartment, an open path FTIR was installed to measure the compartment atmospheric concentration of each agent at a specified level within the compartment. The results from the FTIR represent line-of-sight average values and are reported in Figures 19 to 22 as the integrated area of absorption peaks in the spectral data for each agent. These data present a finer time resolution on agent concentration than the evacuated flask data previously presented. Data for agent release tests involving halon 1301 and the second HCFC-22 test were lost because of instrumentation failures associated with the FTIR. However, the data presented does indicate that a significant amount of mixing occurs very rapidly after release of the agent. In all cases, an initial peak was detected, the magnitude of the peak

SECTION 9. HUMAN EXPOSURE AND ENVIRONMENTAL IMPACT

Table 4. Final agent concentration for 30 minute release test

Compound		Theoretical Average (mg/l)	Measured Average[a] (mg/l)	Final Instantaneous Value Specific Locations		
				Sample I,4 (mg/l)	Sample II,4 (mg/l)	Sample III,4 (mg/l)
FC-116		21.0	11.8 ± 1.2[c]	11.9	10.5	12.9
Halon 1301	1	19.0	8.0 ± 0.8	8.3	7.1	8.6
	2	17.0	7.6 ± 0.5	7.6	7.1	8.0
HCFC-22	1	8.0	4.0 ± 0.4	4.3	3.5	4.2
	2	13.0	5.1 ± 1.0	5.9	5.4	4.0
HFC-227		5.0	3.5 ± 0.1	3.6	3.5	3.5
HFC-32/HFC-125[b]		5.0	3.2 ± 0.1	3.3	3.1	3.3

[a] Average of last sampling flask at each location.
[b] Weighted average values listed for mixture.
[c] Standard deviation of three samples

Table 5. Comparison of final average agent concentration within the compartment to agent concentration at sample probe located 0.31 m from ceiling at 300 s

Compound		Final Average Concentration (mg/l)	Probe 0.31 m at 300 s (mg/l)	Difference (%)
FC-116		11.8	13.7	16.1
Halon 1301	1	8.0	7.9	-1.3
	2	7.6	8.5	11.8
HCFC-22	1	4.0	4.8	20.0
	2	5.1	5.2	2.0
HFC-227		3.5	4.0	14.3
HFC-32/HFC-125		3.2	3.8	18.7

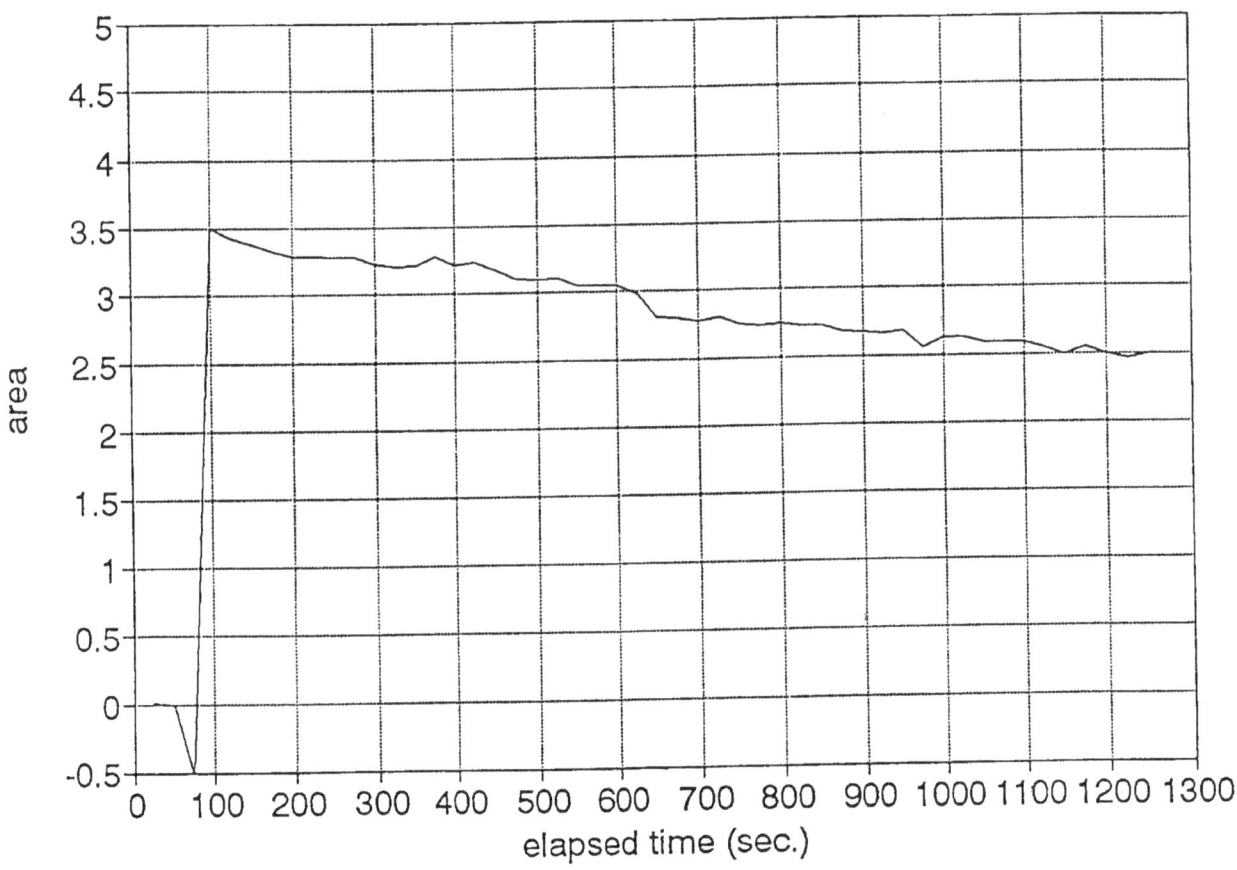

Figure 19. Time dependent agent concentration for agent FC-116 as measured by an open path FTIR located 0.3 m above and slightly off center of the exit tube with measurement beam diagonally across compartment.

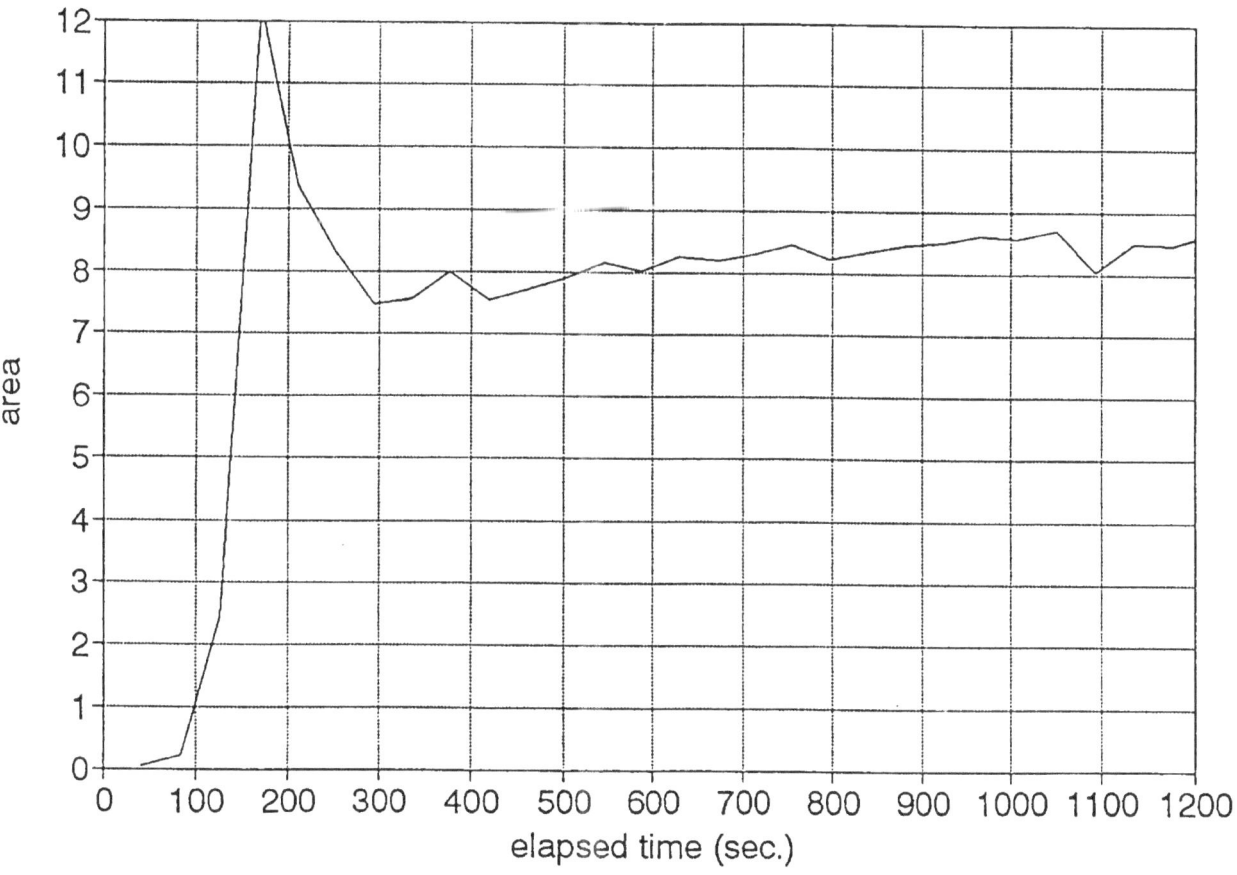

Figure 20. Time dependent agent concentration for agent HCFC-22 as measured by an open path FTIR located 0.3 m above and slightly off center of the exit tube with measurement beam diagonally across compartment.

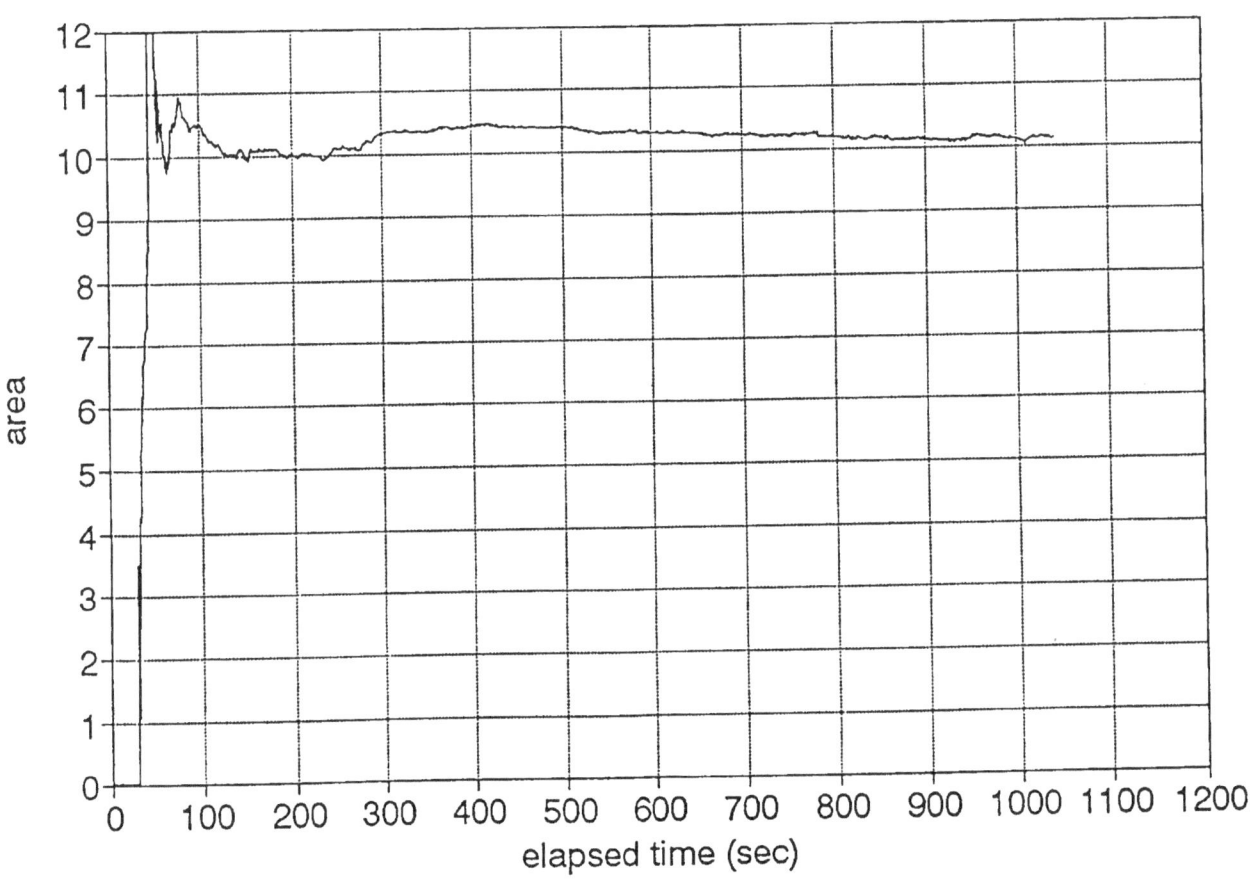

Figure 21. Time dependent agent concentration for agent HCFC-227 as measured by an open path FTIR located 0.3 m above and slightly off center of the exit tube with measurement beam diagonally across compartment.

SECTION 9. HUMAN EXPOSURE AND ENVIRONMENTAL IMPACT 793

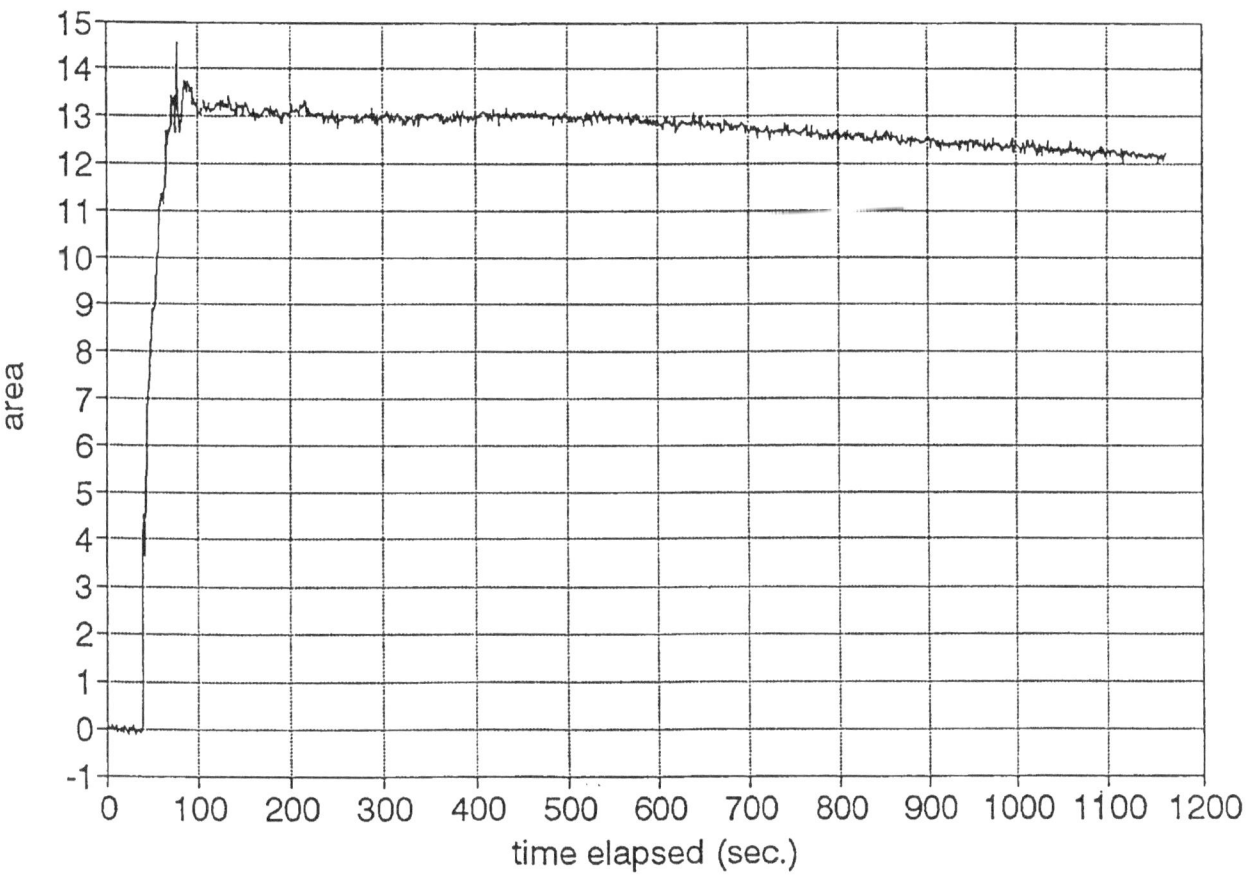

Figure 22. Time dependent agent concentration for agent FC-32/FC-125 as measured by an open path FTIR located 0.3 m above and slightly off center of the exit tube with measurement beam diagonally across compartment.

depending upon the actual position of the FTIR sampling beam relative to the release point, that quickly dissipated, within 60 s. The data for agents HCFC-22, Figure 20, and HCFC-227, Figure 21, show a relatively steady agent concentration after the dissipation of the initial peak. Figure 19, agent FC-116, shows a small initial peak, which may be due to a combination of scan average and short duration of the initial release for this agent, followed by a steady but small decline in agent concentration. These measurements were made with the FTIR beam located approximately 0.3 m above the release point traversing the diagonal of the compartment slightly off center of the exit tube. For agent HFC-32/HFC-125, Figure 22, the FTIR was placed such that the sampling beam traveled along the north wall approximately 0.35 m from the ceiling/wall junction. The data also show a steady but small decline in agent concentration after an initially small peak that lasted less than 60 seconds. Qualitatively, the FTIR data is in agreement with the evacuated flask data previously presented. At the location of the FTIR line-of-sight measurement, immediately after agent release, the agent concentration began to decay to a relatively steady state concentration. Since the FTIR instrument was not calibrated for agent concentration, only relative concentration numbers are reported. Therefore, the FTIR data is not directly comparable to the evacuated flask data. Nevertheless, the relative decay rates calculated from the data in figures 19 and 22, 8.5×10^{-4} and 1.0×10^{-3}, respectively, are within an order of magnitude of those determined by the evacuated flask measurements.

9.1.4 Modeling Results. In an effort to generalize the results of the full scale agent release tests discussed in the previous section, various modeling techniques ranging from an analysis of agent diffusion characteristics to performing 3-D fluid flow calculations using Harwell-Flow3d (CFD 1990) were employed to model the dispersal behavior of halon alternatives escaping from a pressurized storage container into a closed compartment. Techniques for determining diffusion coefficients are available in the literature (Bennett and Myers, 1974). These techniques were used to describe agent diffusion characteristics as a function of temperature and molecular weight. The modeling technique used to simulate agent dispersion in a compartment is to divide the enclosure into a collection of small rectangular boxes or control volumes. Harwell-Flow3d is a member of a class of computer models known as field models that perform this operation. The conditions in each control volume are initially ambient. Agent is then released in several control volumes over time. The resulting flow or exchange of mass, momentum and energy between control volumes is determined so that these three quantities are conserved. The momentum conservation equations are equivalent to Newton's second law of motion and are referred to as the Navier-Stokes equations. The energy conservation equation is equivalent to the first law of thermodynamics. These fluid flow equations are expressed mathematically as a set of simultaneous, non-linear partial differential equations. After being discretized, the resulting finite volume equations are solved iteratively using a variant of Newton's method for computing coupled non-linear algebraic equations. Details of the fluid flow are realized by performing these calculations for each control volume throughout the compartment.

Two main factors affect the dispersion of an agent release within a compartment, **convection** caused by high gas velocities at the exit hole of the pressurized bottle and **molecular diffusion**. These velocities were predicted by Harwell-Flow3d to decrease to a negligible size soon after the bottle had finished evacuating. Figures 23 and 24 show shaded velocity contours in a vertical plane of the bottle evacuation at the end of the bottle evacuation and 65 seconds later. Note how the velocity contours in Figure 24 are small compared to the ones in Figure 23. This was also observed experimentally. Therefore, the significant dispersion occurs during the initial agent release, with diffusion controlling further dispersion.

Bennett and Myers (1974) outline a procedure for determining the diffusion coefficient for a binary mixture (air and the halon alternative in this case) given the temperature, and molecular

Figure 23. Velocity magnitude shaded contours 25 seconds after discharge has begun in an ASTM standard room.

796 SECTION 9. HUMAN EXPOSURE AND ENVIRONMENTAL IMPACT

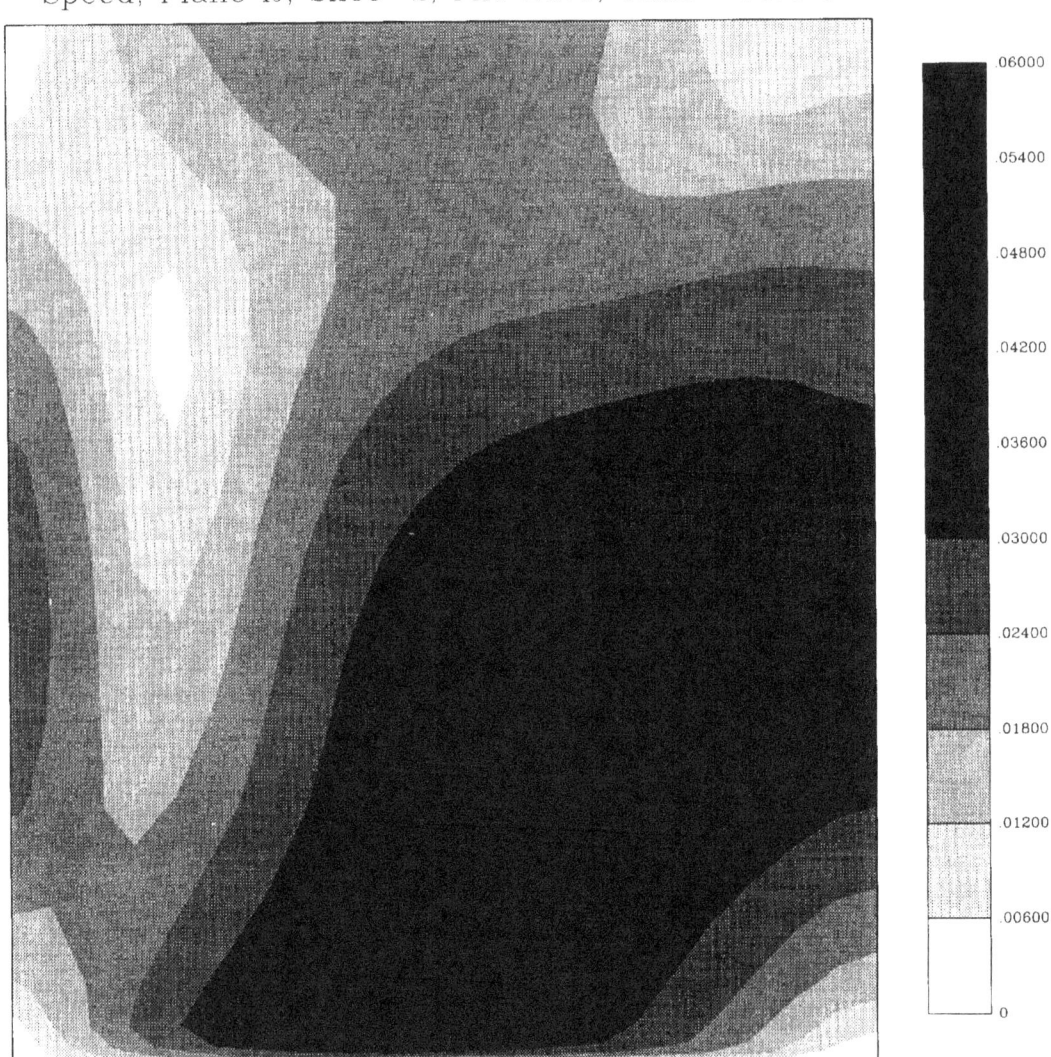

Figure 24. Velocity magnitude shaded contours 90 seconds after discharge has begun in an ASTM standard room.

SECTION 9. HUMAN EXPOSURE AND ENVIRONMENTAL IMPACT

weights of the two gasses. Incorpera and De Witt (1990) list diffusion coefficients for various gases in air ranging from 0.62×10^{-5} m^2/s for naphthalene, $C_{10}H_{10}$, (molecular weight of 132) to 0.16×10^{-4} m^2/s for CO_2 (molecular weight of 46). Since the molecular weights of halon alternatives range from 50 to 240 it is expected that the diffusivity of a halon alternative would be close to this range of 0.62×10^{-5} m^2/s to 0.16×10^{-4} m^2/s. A value of 7.2×10^{-5} was used in the field modeling studies in this report. The distance x that a gas diffuses (assuming the gas is free to diffuse in three dimensions) in $t=600$ seconds for diffusion coefficient within this range can be estimated using the formula $\Delta x = \sqrt{6D\Delta t}$ to be from 0.15 m to 0.24 m. This is a fraction of the room dimensions. Thus, for a quiescent compartment, steady conditions should be reached quickly, with slow changes thereafter. This is consistent with experimentally observed results.

Two scenarios were modeled using Harwell-Flow3d. Both scenarios involved a storage bottle vented upwards in the center of a compartment with dimensions 2.4 m x 3.4 m x 2.4 m. The compartment in both scenarios contained a small leak. In the first scenario, the room was free of obstructions. In the second scenario, the compartment contained two solid baffles in order to partially block the initial flow from the pressurized bottle. Figures 25 and 26 illustrate the results of the simulation after the initial release from the bottle. For the compartment without blockages, the dispersion is relatively uniform throughout. This is consistent with the results from the experimental observations. For the room with blockages, the initial flow is restricted resulting in a more concentrated region contained within the space between the storage bottle and the first baffle. This shows that, for storage areas with exposed structural members along the ceiling or for high bay rack storage areas, vertical obstructions can result in much higher agent concentrations than would normally be anticipated based on the total compartment volume. Care must therefore be exercised in the placement and storage of large quantities of any agent.

9.1.5 Exposure Limits.

The preceding sections indicate that dispersion of these chemicals throughout an unobstructed volume is relatively rapid. Should an accidental discharge occur, the threat to any trapped people would be determined by the toxicity of the fluid. Table 6 presents recommended chronic and acute limits, as available, for the 12 agents. These provide working guidelines (rather than definitive exposure limits) for safe exposures. For many of the agents, little data is available. For each agent, the CAS number provides unique identification of the chemical and much of the material safety data sheet (MSDS) information is summarized in Table 6. In addition, information on human exposure to the agents is included, to the extent such information is available. This information falls in basically four categories: acute exposure (effects of short-term, usually high-level exposure to the chemical), chronic exposure (long-term low-level exposure to the chemical), contact exposure (effects of exposure to the skin or eyes), and carcinogenicity.

For exposure during an accidental release, acute exposure effects are most important. These effects are most often reported by an LC_{50} value. This is the level which caused 50% of the test animals in a given experiment to die during and/or after the stated exposure. In all cases where the toxicity information is known, the levels which cause lethal effects are high (greater than 24 %). However, information for most of the agents is not known. Thus, more information will be required for those agents chosen for further study to place the results in appropriate context.

Chronic exposure data are typically used to determine maximum allowable exposure for workers exposed to chemicals over long periods of time. These limits are established by the Occupational Safety and Health Administration (OSHA), American Conference of Governmental Industrial Hygienists (ACGIH), or National Institute of Occupational Safety and Health (NIOSH). Chronic exposure limits are typically reported by time-weighted averages of maximum allowable concentration over an 8 hour workday or 40 hour work week.

Figure 25. Mass fraction shaded contours 600 seconds after discharge has begun (570 seconds after pressurized bottle has finished evacuating) in an ASTM standard room.

SECTION 9. HUMAN EXPOSURE AND ENVIRONMENTAL IMPACT

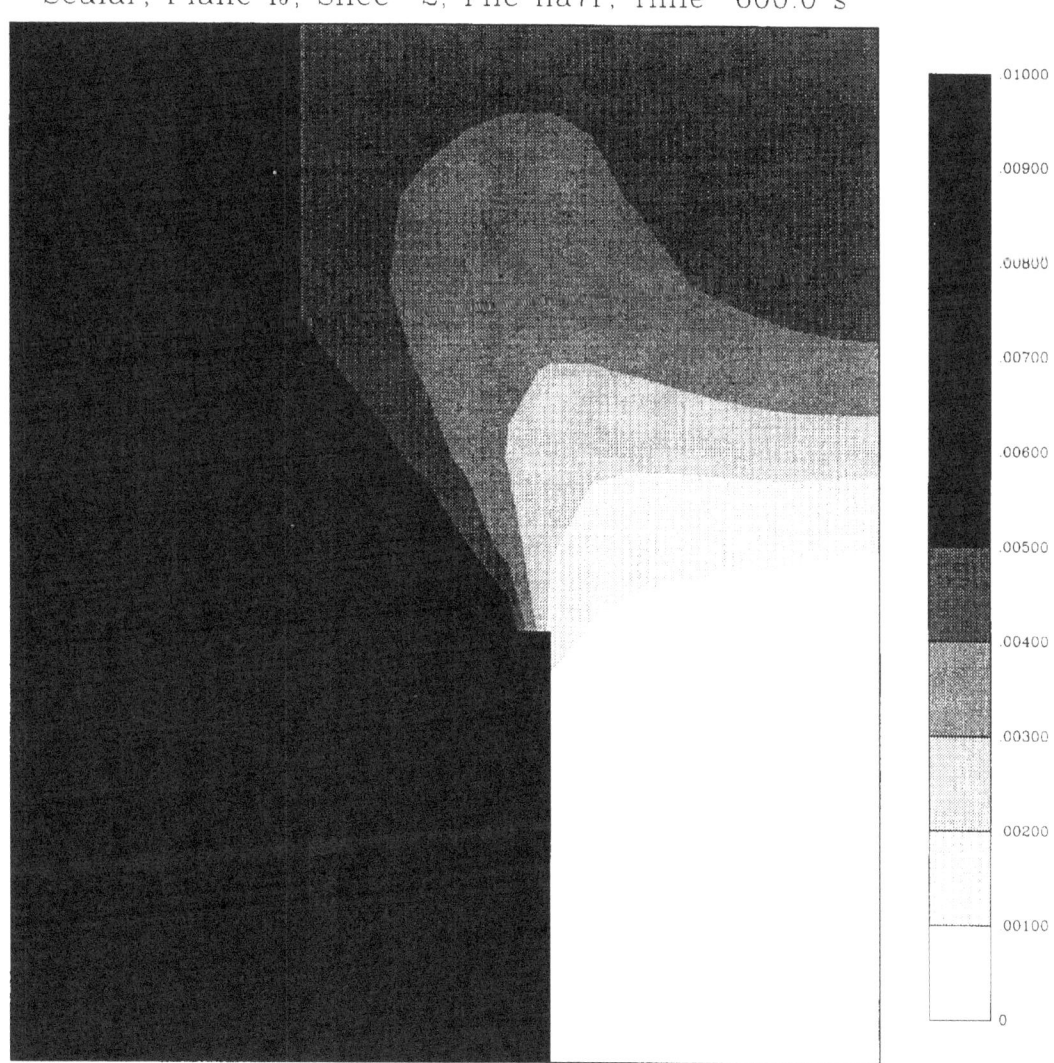

Figure 26. Mass fraction shaded contours 600 seconds after discharge has begun (570 seconds after pressurized bottle has finished evacuating) in an ASTM standard room containing baffles that restrict flow.

SECTION 9. HUMAN EXPOSURE AND ENVIRONMENTAL IMPACT

Table 6. Available Toxicity Information for Candidate Replacement Agents

Agent HFC-32

CAS #	MSDS	Exposure	Toxicity Information
75-10-5	yes 9/91	Acute	No limits established by OSHA[a], ACGIH[b], or NIOSH[c]. Freons in high concentration may cause pulmonary irritation, narcosis, dizziness, incoordination, confusion, nausea, vomiting, tremors, and rarely, coma. When oxygen deficiency has been corrected, these effects appear to reverse.
		Chronic	No limits established by OSHA[a], ACGIH[b], or NIOSH[c].
		Contact	Skin No adverse effects reported. Due to rapid evaporation, frostbite with redness, tingling, pain and numbness may occur. Skin may become hard and white and develop blisters.
			Eyes No adverse effects reported. Due to rapid evaporation, frostbite with redness, pain and blurred vision may occur.
		Carcinogenicity	No data available

a OSHA = Occupational Safety and Health Administration
b ACGIH = American Conference of Governmental Industrial Hygienists.
c NIOSH = National Institute of Occupational Safety and Health

Agent HFC-32/HFC-125

CAS #	MSDS	Exposure	Toxicity Information
75-10-5 354-33-6	yes 9/91 4/89		See above for HFC-32 See below for HFC-125

Agent HFC-227

CAS #	MSDS	Exposure	Toxicity Information
431-89-0	yes 9/90	Acute	No limits established by OSHA, ACGIH, or NIOSH. The toxicological properties of this material have not been determined. Preliminary testing indicates the acute inhalation LC_{50} in rats is greater than 241,000 ppm, 4 hour LC_{50} is greater than 800,000. Stimulants such as epinephrine may induce ventricular fibrillation.
		Chronic	No limits established by OSHA, ACGIH, or NIOSH.
		Contact	Skin No data available
			Eyes No data available
		Carcinogenicity	No data available

SECTION 9. HUMAN EXPOSURE AND ENVIRONMENTAL IMPACT

Table 6. (continued) Available Toxicity Information for Candidate Replacement Agents

Agent HCFC-22

CAS #	MSDS	Exposure	Toxicity Information
75-45-6	yes 10/91	Acute	Relatively non-toxic by inhalation.. Rat LC_{50}: 350,000 ppm (15 min), Mouse LC_{50}: 280,000 ppm (30 min), Dog LC_{50}: 700,000 ppm. Central nervous system depressant, simple asphyxiant. Stimulants such as epinephrine may induce ventricular fibrillation.
		Chronic	1000 ppm (3540 mg/m^3) = OSHA TWA[d], ACGIH TWA, and NIOSH recommended TWA. 1250 ppm (4375 mg/m^3) NIOSH recommended STEL[e]. 500 ppm (1770 mg/m^3) DFG MAK[f] TWA. 1000 ppm (3540 mg/m^3) DFG MAK 60 min peak, momentary value, 3 times/shift.
		Contact	Skin — No adverse effects reported due to the gas. The liquid may cause frostbite; skin could turn red, tingle, and may become hard and white and develop blisters. Pain and numbness may develop.
			Eyes — No adverse effects have been reported from the gas. Due to rapid evaporation, the liquid may cause frostbite with redness, pain and blurred vision.
		Carcinogenicity	No information for humans. Male rats had increased incidences of fibrosarcomas and zymbal-gland tumors. Not found in females.

[d] TWA Time Weighted Average: Time-weighted average concentration for a normal 8 hour workday or 40 hour work week.

[e] STEL Short Term Exposure Limit: 15 min time weighted average exposure which should not be exceeded at any time during a work day even if the 8 hour time weighted average is within established threshold limit value. Exposures at the STEL should not be longer than 15 min and should not be repeated more than 4 times per day with at least 60 min between successive exposures at the STEL.

[f] DFG MAK Maximum concentration values in the work place from the Federal Republic of Germany.

Table 6. (continued) Available Toxicity Information for Candidate Replacement Agents

Agent HFC-134a

CAS #	MSDS	Exposure	Toxicity Information
811-97-2	yes 11/91	Acute	Presently available data are indicative of a low order of toxicity. Nonflammable gas with low water solubility and low chemical reactivity.. LC_{50}(4 hr), rat, >500,000 ppm. Acute symptoms are indicative of central nervous system depression with anesthetic effects induced at sub-lethal levels. The cardiac sensitizing potential is very low (threshold - 75,000 ppm in dogs).
		Chronic	rats, 6 h/day, 5 days/week, 90 days, NOEL = 50,000 ppm.
		Contact	Skin No data available
			Eyes No data available
		Carcinogenicity	rats, one year study, oral doses of 300 mg/kg body weight (in corn oil). After 125 weeks, no carcinogenic potential found.

a NOEL No Observable Effect Level.

Agent FC-116

CAS #	MSDS	Exposure	Toxicity Information
76-16-4	yes 5/92	Acute	No limiting data available. Simple asphyxiant, Inhalation of high concentrations may cause disorientation and narcosis.
		Chronic	No limits established by OSHA, ACGIH, or NIOSH.
		Contact	Skin No adverse effects reported. due to rapid evaporation, liquid may cause frostbite with redness, tingling and pain or numbness. Severe exposures: skin can become hard and white and develop blisters.
			Eyes No adverse effects reported. Due to rapid evaporation, liquid may cause frostbite with redness, pain and blurred vision.
		Carcinogenicity	No data available

SECTION 9. HUMAN EXPOSURE AND ENVIRONMENTAL IMPACT

Table 6. (continued) Available Toxicity Information for Candidate Replacement Agents

Agent HCFC-124

CAS #	MSDS	Exposure	Toxicity Information
2837-89-0	yes 7/91	Acute	Very low acute toxicity. Moderate cardiac sensitizer for dogs exposed to 25,000 ppm plus given epinephrine. In a two week subchronic inhalation study, 100,000 ppm caused mild anesthesia in rats, but recovery occurred in 15 min. Guinea pigs exposed for 30 min, one hour, and two hours at concentration ranging from 9000 to 207,000 ppm had no deaths. Anesthetic effects occurred at concentrations greater than 47,000 ppm, but the guinea pigs recovered following the exposure. High concentrations may cause heart irregularities, unconsciousness or death. Vapors decrease oxygen availability. Frostbite could occur from liquid exposure. The effects in animals from a single exposure by inhalation include central nervous system effects, anesthesia and decreased blood pressure. Repeated exposures increased liver weights, and caused anesthetic effects, irregular respiration, poor coordination, and nonspecific effects such as decreased body weight. However, histopathological evaluation showed no irreversible effects.
		Chronic	No limits established by OSHA, ACGIH, or NIOSH.
		Contact	Skin No data available
			Eyes No data available
		Carcinogenicity	An impurity in FC-124, FC-133a, has been shown to be a potential carcinogen in one published study.

Agent HFC-125

CAS #	MSDS	Exposure	Toxicity Information
354-33-6	yes 4/89	Acute	Not determined. Exposure may be harmful and could cause frostbite. Vapors can cause headache, nausea, giddiness, unconsciousness.
		Chronic	No limits established by OSHA, ACGIH, or NIOSH.
		Contact	Skin No data available
			Eyes No data available
		Carcinogenicity	No data available

Table 6. (continued) Available Toxicity Information for Candidate Replacement Agents

Agent FC-218

CAS #	MSDS	Exposure	Toxicity Information	
76-19-7	yes 10/91	Acute	No limits established by OSHA, ACGIH, or NIOSH. Asphyxia may occur with symptoms such as headache, dizziness, incoordination, dyspnea on mild exertion, sweating, malaise, tremors, convulsive movements, irregular breathing and death.	
		Chronic	No limits established by OSHA, ACGIH, or NIOSH.	
		Contact	Skin	No data available
			Eyes	No data available
		Carcinogenicity	No data available	

Agent FC-31-10

CAS #	MSDS	Exposure	Toxicity Information	
355-25-9	yes 7/92	Acute	The toxicity of this material has not been determined. May be harmful if inhaled, ingested or by skin absorption.	
		Chronic	No limits established by OSHA, ACGIH, or NIOSH.	
		Contact	Skin	No data available
			Eyes	No data available
		Carcinogenicity	No data available	

Agent FC-318

CAS #	MSDS	Exposure	Toxicity Information	
115-25-3	yes 6/92	Acute	No limits established by OSHA, ACGIH, or NIOSH. Simple asphyxiant. Stimulants such as epinephrine may induce ventricular fibrillation.	
		Chronic	No limits established by OSHA, ACGIH, or NIOSH.	
		Contact	Skin	No adverse effects reported. The liquid could cause frostbite accompanied by redness, tingling, pain or numbness. The skin may harden and turn white and develop blisters.
			Eyes	No adverse effects from gas. The liquid could cause frostbite accompanied by redness, pain and blurred vision.
		Carcinogenicity	No data available	

SECTION 9. HUMAN EXPOSURE AND ENVIRONMENTAL IMPACT

Table 6. (continued) Available Toxicity Information for Candidate Replacement Agents

Agent CF_3I

CAS #	MSDS	Exposure	Toxicity Information
2314-97-8	yes 8/90	Acute	The toxicity of this material has not been determined. May have adverse effects if inhaled, ingested or absorbed by the skin.
		Chronic	No limits established by OSHA, ACGIH, or NIOSH.
		Contact	Skin Exposed skin may show signs of frostbite.
			Eyes No data available
		Carcinogenicity	No data available

Table 6 will be updated over the next two years as new information becomes available. A final version will appear in the September, 1995 final report.

9.2 Environmental Requirements for Candidate Replacements for Halon 1301

In addition to being effective in extinguishing fires, a replacement for halon 1301 must not cause greater damage to human health and the environment. Environmental issues include ozone depletion, global warming, water and air pollution. Health and safety effects include acute toxicity from an accidental release (some presented in Section 9.1, above), long term carcinogenic effects from repeated exposures, and fires resulting from the flammability of the chemical.

The U.S. Environmental Protection Agency has established the Significant New Alternative Policy Program (SNAP) under Section 612 of the Clean Air Act to evaluate alternative substances to ozone-depleting chemicals. The program includes a wide range of applications including fire and explosion protection agents, which is the application of interest for this report. The SNAP Program requires a manufacturer to provide specific information in order for EPA to determine the suitability of the agent as a replacement for halon 1301. The information includes not only the environmental/safety parameters but also information pertaining to the amount of substance that would be released and the practicality of using the replacement chemical in terms of availability, cost, and amount required relative to halon 1301. In addition the relevant physical properties of the replacement compound must be included.

Here we consider each of the candidate replacements chemicals relative to the SNAP protocol. We also consider halon 1301 as a reference point and N_2 as a neutral chemical. This review focuses on major issues that might affect the application of a specific chemical rather than on a detailed analysis for each chemical. Information on each of 14 chemicals is contained in a database entitled "Environmental/Health Information on Candidate Replacements for halon 1301" and is also included in hard copy form as Appendix A. The data base is available by request from the editors of this report. The key findings are included in Section 9.2.2 below. Some of the candidate chemicals have already gone through a preliminary screening by EPA, and this information is identified as well.

In the paragraphs below, each of the pertinent elements of the SNAP protocol is described. The remainder of the section discusses the major implications of the tabulated information.

9.2.1 SNAP Protocol. The information required under the SNAP Protocol (Federal Register, 1993) regarding the environmental and health impact of each chemical is described below. The protocol also requires detailed confidential business information and process-specific information, which is not discussed here. Information about each entry is provided below.

1. **Name and Description of Substitute.** The chemical formula and the standard nomenclature for refrigerants are given.

2. **Physical and Chemical Information.** This includes molecular weight, normal boiling point, vapor pressure, and water solubility. This information is pertinent both to the performance of the replacement chemical and to its dispersion in the environment.

3. **Application.** How will the substance be used? In this case the application is for engine nacelle and the dry bay protection. The engine nacelle is the region enclosing the engine. The fire suppression agent would be directed into the nacelle and not into the engine itself. Dry bays are compartments adjacent to fuel cells and can be located in either wing or fuselage areas. They may contain electronic, hydraulic or mechanical equipment. In both of these applications it is important that the agent extinguish the fire rapidly.

4. **Ozone Depletion Potential (ODP).** Stratospheric ozone is critical to blocking ultraviolet (UV) radiation from the sun. An increase in the UV solar radiation will have deleterious effects including increased incidence of skin cancer. The ODP value is a dimensionless number with a value of 1 corresponding to the ODP of CFC-11, CHF_2Cl. The ideal value is zero. The values listed in Appendix A are obtained from The Scientific Assessment of Stratospheric Ozone (Pyle *et al.*, 1991).

5. **Atmospheric Lifetime.** The global warming potential (GWP) of a chemical is determined by its absorption of thermal radiation from the earth and by its tropospheric lifetime relative to the value for carbon dioxide (IPCC, 1990). For the compounds selected here, the atmospheric lifetime provides a good estimate of the GWP. An ideal candidate would have a lifetime of a year or less; whereas, a molecule lasting for 1,000 years would be considered a problem.

6. **Flammability.** The substance should not be flammable at ambient oxygen concentration at atmospheric pressure and temperature. This would create special needs in terms of the storage and use of the chemical. Of course, one would not expect a fire suppressant agent to be flammable.

7. **Toxicity data.** The toxicity data include acute exposure leading to death, the chronic effects of a long term, low level exposure, cardiac sensitization such as an increase in the pulse or irregular heartbeats, and the tendency to cause cancer or to have a mutagenic effect. Instead of determining the absolute health impact of a chemical, it is often easier to compare its impact with that of halon 1301.

8. **Release.** The expected amount of chemical released per event. The amount released for an engine or dry bay fire is on the order of a fraction of a kg for halon 1301.

SECTION 9. HUMAN EXPOSURE AND ENVIRONMENTAL IMPACT

9. **Replacement Ratio.** This ratio is based on the volume of an agent (as a saturated liquid @ 25 °C) relative to the volume of halon 1301 necessary to suppress various flame arrangements. Our value corresponds to the average of spray burner and detonation tube tests discussed in Section 4. This is thought to be the most appropriate number for dry bays. A different estimate for the replacement ratio is obtained for engine nacelles and the resulting values (see Section 4) are as much as 50% greater than the dry bay values.

10. **Availability.** The issue is whether the chemical is currently (as of September 30, 1993) available in the amount required for replacing halon 1301.

11. **Cost.** The cost is a variable, which fluctuates with the market. The tabulated values are our best estimates as of September 30, 1993 based on the purchase of 909 kg (2000 lbs) of the agent.

12. **Required Technological Changes.** This refers to changes in the method for dispensing the agent relative to halon 1301.

9.2.2 Discussion of Key Results in Database.
A summary of the key issues for each agent is included in Table 7, with details in Appendix A. Important aspects of the information in the database are discussed in the sections below.

9.2.2.1 Physical/Chemical Information.
The database includes information for 16 different compounds. It is helpful to divide the compounds into various classes. Two chemicals stand out from the others: N_2 and $NaHCO_3$. Nitrogen was chosen to provide test data for a gas with no expected chemical effect on a fire and no environmental impact since the earth's atmosphere is 80% nitrogen. The $NaHCO_3$ dry powder releases CO_2 when heated.

The other compounds involve carbon bonded to halogens (fluorine, chlorine, bromine, and iodine), to hydrogen, and/or to other carbon atoms. The bond dissociation energy (BDE) decreases monotonically with increasing atomic weight of the halogen for the carbon halogen bonds in CF_3X from a value of 546 kJ per mole for C-F (McMillan *et al.*, 1982), to 331 for C-Cl, to 301 for C-Br, to 223 for C-I. The last three BDE's were computed by the method described in Section 6.1.3.1.5 (Nyden, 1993). The C-H BDE in CF_3H is 447 kJ/mole (McMillan *et al.*, 1982) and the C-Cl BDE in CHF_2Cl is about 330 kJ/mole. It is convenient to classify fire extinguishing agents with regard to the bond most easily broken. We have the following classes in order from weakest to strongest:

C-I bond (CF_3I)
C-Br bond (CF_3Br)
C-Cl bond (HCFC-22, HCFC-124)
C-H bond (HFC-32, HFC-125, HFC-32/HFC-125, HFC-227ea, HFC-236fa, HFC-134a)
C-F bond (FC-116, FC-218, FC-3110, FC-318)

As discussed below, the smaller BDE correlates with a shorter atmospheric lifetime; for example, the estimated lifetime of CF_3I is about 2 weeks compared to an estimated 10,000 years for FC-116 (Wuebbles, 1993).

Another qualitative trend is the decreasing vapor pressure and increasing boiling point with increasing molecular weight. For example the normal boiling point is -78 °C for FC-116 (C_2F_6) compared to -2 °C for FC-3110 (C_4F_{10}). The vaporization characteristics are important for the two phase dispersal as the fluid leaves the nozzle.

Table 7. Summary of Issues Regarding Candidate Chemicals

Agent	Comments
halon 1301	Not acceptable because of high ozone depletion potential
CF_3I	ODP not known; toxicity and corrosivity not known
sodium bicarbonate	Corrosive to aluminum; technology for rapidly dispersing the powder must be developed
perfluorinated compounds, (FC-116, FC-218, FC-318, FC-3110)	long atmospheric lifetime; high price for FC-318
HCFC-22, HCFC-124	Both on SNAP short list; however, finite ozone depletion potential and cardiac sensitivity
HFC-32	Not acceptable because of flammability
HFC-125, HFC-227	Both on SNAP short list; slight cardiac sensitivity
HFC-32/HFC-125	High HF production from flame; cardiac sensitivity
HFC-134a	On SNAP short list; high HF production from flame, cardiac sensitivity
HFC-236fa	Possible cardiac sensitivity; more difficult to disperse because of high boiling point

Because of their high vapor pressures, these chemicals are not expected to persist as a liquid pool or dissolved in water. The water solubility for most of the compounds is small. The most soluble is HCFC-22 (CHF_2Cl) with a water solubility of 3.3 kg/m^3 followed by CF_3Br with a value of 0.3 kg/m^3. The available data indicate that the other compounds are less soluble. Even the moderate solubility of HCFC-22 is not a water pollution issue for a release in an engine nacelle or dry bay.

9.2.2.2 Ozone Depletion Potential. As can be seen in the results tabulated in Appendix A, the ozone depletion potential of most of the compounds is zero. The exceptions are CF_3Br, which is being proscribed because of its high ozone depletion potential of 16; CF_3I, which is currently being evaluated, and the two chlorine containing molecules, HCFC-22 and HCFC-124, with ozone depletion potentials of 0.055 and 0.022, respectively. These values are considered small enough relative to what they replace that both of these chemicals are on the short list of proposed acceptable alternatives under SNAP Program.

9.2.2.3 Atmospheric Lifetime. As discussed above, the tropospheric lifetime is used as a surrogate for the global warming potential. Based on the heuristic bond energy analysis given above, one would expect CF_3I to have the shortest atmospheric lifetime since it has the smallest bond energy. This expectation is in qualitative agreement with the statement in Section 5 that the absorption spectra of CF_3I is shifted to the red relative to the other CF_3X's. Work in progress also indicates a very short lifetime of about 2 weeks (Wuebbles, 1993). Since $NaHCO_3$ releases CO_2 when heated, its impact on the environment is determined by the CO_2 life-cycle and is not considered to be an issue. Chlorine containing molecules are next in increasing atmospheric lifetimes followed by the hydrogen containing molecules. The lifetime of the perfluorinated compounds is the longest, with Wuebbles (1993) estimating 50,000 years for CF_4. Because the five candidate perfluorinated compounds considered here have C-C bonds, which are weaker than the C-F bond, the lifetime is expected to be less than 50,000 years but may still be on the order of tens of thousands of years. This long lifetime is a serious issue with regard to their suitability as candidate replacement chemicals.

9.2.2.4 Flammability. The chemical HFC-32 (CH_2F_2) is flammable at an ambient oxygen concentration. In a mixture with HFC-125, HFC-32 is not flammable. While the compounds are not flammable, one should note that the increased number of C-C and C-H bonds will increase the heat release of these molecules in a flame environment.

9.2.2.5 Replacement Ratio. In all cases except $NaHCO_3$ more material will be required than for the currently used suppression agent CF_3Br; however, in the worst case only about twice as much is required. While this is a narrow range, it could still be important because of the weight and volume limitations on an aircraft. The replacement ratio computed here is on a volumetric basis. The corresponding mass ratio, that is the mass of an agent relative to the mass of halon 1301 necessary to suppress various flame arrangements, can be determined from the tabulated results and the densities given in Sections 2 and 4 of this report.

9.2.2.6 SNAP Alternative. There are five compounds from the above list that are on EPA's short list as proposed acceptable alternatives under the SNAP Program for total flooding of unoccupied areas. These are HFC-125 (C_2HF_5), HFC-227ea (C_3HF_7), HCFC-22 (CHF_2Cl), HFC-134a ($C_2H_2F_4$), and HCFC-124 (C_2HF_4Cl). Many of the other chemicals in the database we are considering are similar to these. The exceptions are the perfluorinated compounds which have a significantly longer atmospheric lifetime than any of the other chemicals.

9.2.2.7 Required Technological Changes. For each agent there will likely be some changes required relative to the current halon 1301 technology in terms of size of container, material for container, valve type, lubricant, etc. However, for one of these, $NaHCO_3$, a completely different dispersal system must be developed, since the $NaHCO_3$ is in the form of a powder. To be effective, the powder must be distributed essentially as fast as a gas.

9.2.2.8 Availability. The issue of availability is a critical one in the case of iodotrifluoromethane, CF_3I, which, as of this writing, is made in amounts of a few kg per batch. One company (DuBoisson, 1993) uses the so-called Hunsdiecker reaction (Banks, 1971). This involves the careful pyrolysis of an anhydrous mixture of silver trifluoroacetate and iodine:

$$CF_3CO_2Ag + I_2 \rightarrow CF_3I(>90\%) + CO_2 + AgI \qquad (1)$$

The reaction is initiated by gently heating the reagents to about 100 °C. Once initiated the reaction is exothermic and the dissipation of the heat is one of the difficulties in scaling up this process above a few kilogram batch. Another difficulty is the recovery of the AgI from the vessel for reprocessing. The current technology involves collecting the CF_3I in cooled tubes connected to the reactor vessel and breaking the reactor vessel each time to recover the AgI.

Another company (Newhouse, 1993) has just begun manufacturing CF$_3$I using a 200 l reactor with a batch product of about 25 kg. They claim that there is not a major impediment to scaling up the operation. However, until this is demonstrated the availability of CF$_3$I will remain an issue.

9.2.2.9 Combustion Products. One of the major unwanted products of combustion is HF. Based on cup burner experiments, Appendix A shows that there is more than a factor of 10 variation in the amount of HF released per gram fuel consumed with the two poorest performers, HFC-32/HCF-125 and HFC134a. The HF production of the other compounds is a factor of two or more less than these compounds. For the CF$_3$I there will also be the issue of the production of HI and I$_2$.

9.2.2.10 Price. There is a tremendous range in the price from as low as $3.50 per kg for HCFC-22 to $210 per kg for FC-318 and $330 per kg for CF$_3$I. A major factor in the high price is the limited production of some of the compounds. The cost is likely to drop significantly with greater production.

9.2.2.11 Toxicity. These chemicals are intended for use in areas external to the inhabited areas of the airplane. Therefore, people will not be exposed to the suppressant during fire suppression. However, there is a chance of accidental exposure during the installation and maintenance of the system. In such a case, safety considerations following rapid release of the two-phase flow include:

- low temperatures arising from the flow expansion, and
- asphyxiation from release into a confined space.

Chronic effects from repeated low level exposures should also be considered.

Our toxicity data are abstracted from the Air Force literature survey (Jepson, 1993). The perfluorinated chemicals, which are apparently unreacted in the body, appear to be the least toxic of the candidate chemicals and compare favorably with halon 1301 in acute testing. On the other hand, the most toxic appear to be the chlorine-containing chemicals (HCFC-22 and HCFC-124) because of their elevated levels of cardiac sensitization relative to the other chemicals. There is also a growing popular concern about the toxicity of any chlorine-containing chemical. It is noteworthy that, even so, these chemicals are on the short list for proposed alternatives under the SNAP Program. The chemical HFC-134a, C$_2$H$_2$F$_4$, has about the same cardiac sensitization as halon 1301 and has been shown to undergo a metabolism similar to anesthetic agent halothane. This is of some concern since repeated low level exposures of personnel to the anesthetic agent halothane can produce viral-like hepatitis. The perfluorinated halocarbons do not produce the viral-like behavior because they are biologically inert. Toxicity data for CF$_3$I are just beginning to be generated at this time.

Little is known about the mutagenic/carcinogenic potential of the agents for humans. FC-318, FC-218, and HCFC-22 have all demonstrated some mutagenic activity in an *in vitro* system. Only HCFC-22 has been tested in a mammalian system; based on the results it is classified as Group 3 (the agent is not classifiable as to its carcinogenicity in humans).

9.3 Conclusions

- Experimental measurement, hand calculations, and 3-D fluid flow modeling of an accidental release of a halon alternative show that a reasonably uniform concentration of agent is obtained in a compartment shortly after release of the agent. Thus, simple ideal gas calculations of agent concentrations should suffice for estimating agent concentration following an accidental release.

- For compartments which include baffles to the gas flow, the diffusion controlled flow after agent release may confine the released agent within the baffles near the release point of the agent.

- As of this report, only five candidate agents have been designated as acceptable alternative chemicals to halon 1301. These are: HCFC-124, HCFC-22, HFC-134A, HFC-227, and HFC-125.

9.4 References

Banks, R.E., *Fluorocarbons and their Derivatives*, Elsevier Publishing Co., Inc., New York, 1971.

Barrow, G.M., *Physical Chemistry*, McGraw-Hill, Inc., New York, 1961.

Bennett, C.O. and Myers, J.E., *Momentum, Heat, and Mass Transfer*, McGraw-Hill, Inc., New York, second edition, 1974.

CFD Department, AEA Industrial Technology, Harwell Laboratory, Oxfordshire, United Kingdom. *HARWELL-FLOW3D Release 2.3: User Manual*, 1990.

DuBoisson, R., PCR, Gainesville, Florida, personal communication, 1993.

Federal Register, Volume 58, 90, pp. 28102-28104, May 12, 1993.

Incorpera, F.P. and De Witt, D. P., *Fundamentals of Heat and Mass Transfer*, John Wiley and Sons, New York, third edition, 1990.

Intergovernmental Panel on Climate Change (IPCC): Climate Change; The IPCC Scientific Assessment, Cambridge University Press, Cambridge, UK, 1990.

Jepson, G.W., Memorandum for Capt. Jepson, WL/FIVS (Mr. Bennett), May 21, 1993.

McMillan, D.F., and Golden, D.M., "Hydrocarbon Bond Dissociation Energies," *Annual Review of Physical Chemistry* 33 ,493 (1982)

Newhouse, S., Pacific Scientific, Duarte, CA, personal communication, 1993.

Nyden, M., National Institute of Standards and Technology, personal communications,1993.

Pyle, J.A., Solomon, S., Wuebbles, D., and Zvenigorodsky, S., "Ozone Depletion and Chlorine Loading Potentials," in World Meteorological Organization, Global Ozone Research Monitoring Project - Report No. 25, *Scientific Assessment of Ozone Depletion: 1991*, 1991.

Wuebbles, D.J., Lawrence Livermore National Laboratory, personal communication, 1993.

SECTION 9. HUMAN EXPOSURE AND ENVIRONMENTAL IMPACT

Appendix A. Halon Replacement Properties

```
                  HALON Replacement Chemical Properties Table
Date Recorded: 09/29/93
Formula: CF3Br                   Designation: Halon 1301
      Molecular Weight: 148.9 kg/kmole
        Normal Boiling: 215 K
        Vapor Pressure:   1610 kPa @    298 K
      Water Solubility: 0.3000 kg/m3 @ 101 kPa
        Ozone Depletion
             Potential: 16
       Atmospheric Life
            Time (yrs): 110
          Flammability: NO
      Replacement Ratio: 1.0 based on HALON 1301
      SNAP Alternative: -
            Application: ENGINE NACELLE/DRY BAY
           Availability: UNLIKELY AFTER 1995
                  Price: $61.60 per kg @ 13 kg
Required Technological
               Changes: NONE

Environmental--Release: ABOUT 4 KG PER INCIDENT
   Combustion Products: 0.08 G HF/G C3H8 BURNED
Toxicity---------Acute:
               Cardiac:
          Carcinogenic:
              Mutagenic:
              Pyrolysis:
```

SECTION 9. HUMAN EXPOSURE AND ENVIRONMENTAL IMPACT

```
               HALON Replacement Chemical Properties Table
Date Recorded: 10/01/93
Formula: CF3I            Designation: Iodotrifluoromethane
     Molecular Weight: 196.0 kg/kmole
       Normal Boiling: 251 K
        Vapor Pressure:
      Water Solubility:
       Ozone Depletion
             Potential:
      Atmospheric Life
           Time (yrs): 0.04
          Flammability:
     Replacement Ratio: 1.2 based on HALON 1301
     SNAP Alternative:
           Application: ENGINE NACELLE/DRY BAY
          Availability: CURRENT CAPACITY ABOUT 25 KG PER BATCH
                 Price: $330.00 per kg @ 25 kg
Required Technological
               Changes:

Environmental--Release: ABOUT 4 KG PER INCIDENT
   Combustion Products:
Toxicity---------Acute:
               Cardiac:
          Carcinogenic:
             Mutagenic:
             Pyrolysis:
```

SECTION 9. HUMAN EXPOSURE AND ENVIRONMENTAL IMPACT

```
                    HALON Replacement Chemical Properties Table
Date Recorded: 10/01/93
Formula: NaHCO3                    Designation: Sodium Bicarbonate
    Molecular Weight: 84.0 kg/kmole
      Normal Boiling:
      Vapor Pressure:
     Water Solubility: 69.000 kg/m3 @ 101 kPa
      Ozone Depletion
           Potential: 0.0
     Atmospheric Life
          Time (yrs):
         Flammability: NO
    Replacement Ratio: 0.8 based on HALON 1301
     SNAP Alternative:
          Application: ENGINE NACELLE/DRY BAY
         Availability: YES
                Price: $2.00 per kg @ 1000 kg
Required Technological
              Changes: MAJOR CHANGES REQUIRED TO RAPIDLY DISPERSE A POWDER
Environmental--Release: ABOUT 4 KG PER INCIDENT
   Combustion Products: NaOH
Toxicity---------Acute: NOT A CONCERN
              Cardiac: NOT A CONCERN
         Carcinogenic: NOT A CONCERN
           Mutagenic: NOT A CONCERN
            Pyrolysis: NOT A CONCERN
```

SECTION 9. HUMAN EXPOSURE AND ENVIRONMENTAL IMPACT

```
              HALON Replacement Chemical Properties Table
Date Recorded: 10/01/93
Formula: C2F6                  Designation: FC-116
      Molecular Weight: 138.0 kg/kmole
         Normal Boiling: 195 K
         Vapor Pressure: 3070 kPa @   293 K
        Water Solubility:
        Ozone Depletion
              Potential: 0.0
        Atmospheric Life
             Time (yrs): 10000
           Flammability: NO
      Replacement Ratio: 3.2 based on HALON 1301
       SNAP Alternative:
            Application: ENGINE NACELLE/DRY BAY
           Availability:
                  Price: $37.30 per kg @ 450 kg
Required Technological
                Changes:

Environmental--Release: ABOUT 4 KG PER INCIDENT
    Combustion Products: 0.7 G HF/G C3H8 BURNED
Toxicity---------Acute: RAT, 1 hr @ 800,000 ppm - initial hyperactivity
                        followed by hypoactivity, hyperemia, and closed
                        eyes.
                Cardiac: dog, up to 600,000 ppm - no sensitization
           Carcinogenic:
              Mutagenic: not mutagenic in exposed E. coli.
              Pyrolysis: rat, exposure of pyrolysis products from acetone
                         fire resulted in dyspnea, salivation, lacrimation,
                         and gasping.  Substantial weight loss.
```

SECTION 9. HUMAN EXPOSURE AND ENVIRONMENTAL IMPACT

```
                  HALON Replacement Chemical Properties Table
Date Recorded: 10/01/93
Formula: C3F8                    Designation: FC-218
    Molecular Weight: 188.0 kg/kmole
      Normal Boiling: 236 K
      Vapor Pressure:    880 kPa @   298 K
    Water Solubility:
    Ozone Depletion
           Potential: 0.0
    Atmospheric Life
          Time (yrs): 10000
        Flammability: NO
   Replacement Ratio: 1.6 based on HALON 1301
    SNAP Alternative:
         Application: ENGINE NACELLE/DRY BAY
        Availability:
               Price: $39.60 per kg @ 654 kg
Required Technological
             Changes:

Environmental--Release: ABOUT 4 KG PER INCIDENT
   Combustion Products: 1.1 G HF/G C3H8 BURNED
Toxicity---------Acute: rat, 1 hr @ 800,000 ppm - initial hyperactivity
                        followed by hypoactivity, hyperemia, and closed
                        eyes.
             Cardiac:
        Carcinogenic:
           Mutagenic: Biochemical mutants were produced in E. coli.
           Pyrolysis:
```

HALON Replacement Chemical Properties Table

```
Date Recorded: 10/01/93
Formula: C4F10                    Designation: FC-31-10
   Molecular Weight: 238.0 kg/kmole
     Normal Boiling: 271 K
     Vapor Pressure:    265 kPa @    298 K
    Water Solubility:
    Ozone Depletion
          Potential: 0.0
    Atmospheric Life
         Time (yrs): 10000
        Flammability: NO
   Replacement Ratio: 1.7 based on HALON 1301
   SNAP Alternative:
        Application: ENGINE NACELLE/DRY BAY
       Availability:
              Price: $33.00 per kg @ 600 kg
Required Technological
            Changes:

Environmental--Release: ABOUT 4 KG PER INCIDENT
   Combustion Products: 1.2 KG HF/G C3H8 BURNED
Toxicity---------Acute: rat, 16 hrs @ 800,000 ppm no effect
              Cardiac: dog, no cardiac sensitization noted at up to
                       400,000 ppm
         Carcinogenic:
            Mutagenic:
            Pyrolysis:
```

SECTION 9. HUMAN EXPOSURE AND ENVIRONMENTAL IMPACT

```
                    HALON Replacement Chemical Properties Table
Date Recorded: 10/01/93
Formula: CYCLO C4F8              Designation: FC-318
     Molecular Weight: 200.0 kg/kmole
        Normal Boiling: 266 K
         Vapor Pressure:    310 kPa @    298 K
       Water Solubility:
        Ozone Depletion
              Potential: 0.0
       Atmospheric Life
            Time (yrs): 10000
           Flammability: NO
      Replacement Ratio: 1.7 based on HALON 1301
       SNAP Alternative:
            Application: ENGINE NACELLE/DRY BAY
           Availability: LIMITED PRODUCTION
                  Price: $209.00 per kg @ 45 LBS
Required Technological
                Changes:

Environmental--Release: ABOUT 4 KG PER INCIDENT
    Combustion Products:
Toxicity---------Acute: rat, 4 hr @ 800,000 ppm, no deaths, signs of
                         irritation
               Cardiac: mouse, up to 400,000 ppm alone - no arrhythmias;
                         200,000 ppm + epinephrine - arrhythmias
          Carcinogenic:
             Mutagenic: 5/1300 visible mutation (0.38%) compared to 0.008%
                         spontaneous control rate; sex-linked recessive
                         lethal mutation was not significant.
              Pyrolysis:
```

SECTION 9. HUMAN EXPOSURE AND ENVIRONMENTAL IMPACT

```
            HALON Replacement Chemical Properties Table
Date Recorded: 10/01/93
Formula: CH2F2              Designation: HFC-32
    Molecular Weight: 52.0 kg/kmole
      Normal Boiling: 221 K
      Vapor Pressure:
     Water Solubility:
      Ozone Depletion
            Potential: 0.0
     Atmospheric Life
          Time (yrs): 7
         Flammability: YES
    Replacement Ratio:
     SNAP Alternative:
          Application: ENGINE NACELLE/DRY BAY
         Availability:
                Price:
Required Technological
              Changes:

Environmental--Release: ABOUT 4 KG PER INCIDENT
   Combustion Products:
Toxicity---------Acute: rat, 4 hr @ 111,000 to 760,000 ppm - signs of
                        lethargy, loss of mobility, spasms and gnawing of
                        cage during exposure.  No abnormal signs 30 min to
                        14 days after exposure.
              Cardiac: 1/12 dogs sensitized @ 250,000 ppm; same dog was
                        normal after 200,000 ppm.
         Carcinogenic:
            Mutagenic: Ames assay - negative
            Pyrolysis: no information
```

SECTION 9. HUMAN EXPOSURE AND ENVIRONMENTAL IMPACT

```
                  HALON Replacement Chemical Properties Table
Date Recorded: 10/01/93
Formula: CH2F2/C2HF5          Designation: HFC-32/HFC-125
    Molecular Weight: kg/kmole
      Normal Boiling: 220 K
      Vapor Pressure: 1670 kPa @    298 K
     Water Solubility:
      Ozone Depletion
            Potential: 0.0
     Atmospheric Life
          Time (yrs):
         Flammability: NO
    Replacement Ratio: 2.5 based on HALON 1301
     SNAP Alternative:
          Application: ENGINE NACELLE/DRY BAY
         Availability:
                Price: $33.00 per kg @ 680 kg
Required Technological
              Changes:

Environmental--Release: ABOUT 4 KG PER INCIDENT
   Combustion Products: 6.2 G HF/G C3H8 BURNED
Toxicity---------Acute:
              Cardiac:
         Carcinogenic:
            Mutagenic:
            Pyrolysis:
```

SECTION 9. HUMAN EXPOSURE AND ENVIRONMENTAL IMPACT

HALON Replacement Chemical Properties Table

Date Recorded: 10/01/93
Formula: C2HF5   Designation: HFC-125
    Molecular Weight: 120.0 kg/kmole
      Normal Boiling: 224 K
      Vapor Pressure: 1380 kPa @   298 K
     Water Solubility:
      Ozone Depletion
             Potential: 0.0
     Atmospheric Life
            Time (yrs): 40.5
         Flammability: NO
    Replacement Ratio: 1.8 based on HALON 1301
     SNAP Alternative: +
           Application: ENGINE NACELLE/DRY BAY
          Availability:
                 Price: $31.46 per kg @ 490 kg
Required Technological
              Changes:

Environmental--Release: ABOUT 4 KG PER INCIDENT
   Combustion Products: 2.1 KG HF/G C3H8 BURNED
Toxicity---------Acute:
              Cardiac:
         Carcinogenic:
            Mutagenic: Ames assay - negative, human lymphocyte - negative,
                       CHO cell assay - negative
            Pyrolysis:

SECTION 9. HUMAN EXPOSURE AND ENVIRONMENTAL IMPACT

```
                HALON Replacement Chemical Properties Table
Date Recorded: 10/01/93
Formula: C3HF7                    Designation: HFC-227
    Molecular Weight: 170.0 kg/kmole
      Normal Boiling: 257 K
      Vapor Pressure:
     Water Solubility:
      Ozone Depletion
            Potential: 0.0
      Atmospheric Life
           Time (yrs): 13
         Flammability: NO
    Replacement Ratio: 1.6 based on HALON 1301
     SNAP Alternative: +
          Application: ENGINE NACELLE/DRY BAY
         Availability:
                Price: $49.50 per kg @ 654 kg
Required Technological
              Changes:

Environmental--Release: ABOUT 4 KG PER INCIDENT
   Combustion Products: 1.7 G HF/G C3H8 BURNED
Toxicity---------Acute: rats, 4 hrs @25,000 or 53,000 ppm - no deaths,
                        irreg. breathing, slight lacrimation, red ears,
                        no post exposure effects.
              Cardiac:
         Carcinogenic:
            Mutagenic:
            Pyrolysis:
```

```
                     HALON Replacement Chemical Properties Table
Date Recorded: 10/01/93
Formula: C3H2F6                     Designation: HFC-236FA
    Molecular Weight: 152.0 kg/kmole
      Normal Boiling: 272 K
      Vapor Pressure:    270 kPa @    298 K
    Water Solubility:
    Ozone Depletion
           Potential: 0.0
    Atmospheric Life
         Time (yrs):
        Flammability: NO
   Replacement Ratio: 1.6 based on HALON 1301
    SNAP Alternative:
         Application: ENGINE NACELLE/DRY BAY
        Availability:
               Price:
Required Technological
             Changes·

Environmental--Release: ABOUT 4 KG PER INCIDENT
   Combustion Products: 1.9 G HF/G C3H8 BURNED
Toxicity---------Acute:
              Cardiac:
         Carcinogenic:
            Mutagenic:
            Pyrolysis:
```

SECTION 9. HUMAN EXPOSURE AND ENVIRONMENTAL IMPACT

```
                  HALON Replacement Chemical Properties Table
Date Recorded: 10/01/93
Formula: C2H2F4                      Designation: HFC-134A
      Molecular Weight: 102.0 kg/kmole
        Normal Boiling: 247 K
        Vapor Pressure:    670 kPa @    298 K
       Water Solubility:
       Ozone Depletion
             Potential: 0.0
       Atmospheric Life
            Time (yrs): 15.6
          Flammability: NO
     Replacement Ratio: 1.9 based on HALON 1301
      SNAP Alternative: +
           Application: ENGINE NACELLE/DRY BAY
          Availability:
                 Price: $11.55 per kg @ 680 kg
Required Technological
               Changes:

Environmental--Release: ABOUT 4 KG PER INCIDENT
   Combustion Products: 3.9 G HF/G C3H8 BURNED
Toxicity---------Acute: rat, 4 hr LC50 > 500,000 ppm
               Cardiac: Threshold level - 75,000 ppm
          Carcinogenic:
             Mutagenic: Ames assay - negative, clastogenicity - negative;
                        rat cytogenetics up to 50,000 ppm negative, mouse
                        dominant lethal up to 50,000 ppm - negative.
             Pyrolysis:
```

SECTION 9. HUMAN EXPOSURE AND ENVIRONMENTAL IMPACT

```
             HALON Replacement Chemical Properties Table
Date Recorded: 10/01/93
Formula: CHF2CL              Designation: HCFC-22
    Molecular Weight:  87.0 kg/kmole
      Normal Boiling:  232 K
      Vapor Pressure:   1050 kPa @   298 K
    Water Solubility: 3.3000 kg/m3 @ 101 kPa
      Ozone Depletion
           Potential: 0.055
     Atmospheric Life
          Time (yrs): 15.8
        Flammability: NO
   Replacement Ratio: 2.2 based on HALON 1301
    SNAP Alternative: +
         Application: ENGINE NACELLE/DRY BAY
        Availability:
               Price: $3.50 per kg @ 680 kg
Required Technological
            Changes:

Environmental--Release: ABOUT 4 KG PER INCIDENT
   Combustion Products: 1.3 G HF/G C3H8 BURNED
Toxicity---------Acute: rats, guinea pigs; 2 hr exposure - 75,00 - 100,000
                        ppm -excitation and/or ataxia; 200,000 ppm - CNS
                        depression; 300,000 - 400,000 ppm - death.
              Cardiac: dogs and monkeys; 50,000 - 100,000 ppm caused early
                        respiratory depression, bronchoconstriction, tachy-
                        cardia, myocardial depression and hypotension.
         Carcinogenic: Inadequate evidence of carcinogenicity in humans;
                        some evidence of tumors in male mice exposed for a
                        long period of time.
            Mutagenic: Ames assay, BHK21 cell transformation - positive
                        results; no induction of unscheduled DNA synthesis
                        in human heteroploid EUE cells treated with a 20 mM
                        soln. generated at 500 ml/min (1:1 air) in presence
                        or absence of S10.
            Pyrolysis: Combustion products highly toxic
```

SECTION 9. HUMAN EXPOSURE AND ENVIRONMENTAL IMPACT

```
                 HALON Replacement Chemical Properties Table
Date Recorded: 10/01/93
Formula: C2HF4Cl                    Designation: HCFC124
    Molecular Weight: 137.0 kg/kmole
      Normal Boiling: 260 K
      Vapor Pressure:    380 kPa @    298 K
     Water Solubility:
      Ozone Depletion
            Potential: 0.022
      Atmospheric Life
          Time (yrs): 6.9
         Flammability: NO
    Replacement Ratio: 1.6 based on HALON 1301
     SNAP Alternative:  +
          Application: ENGINE NACELLE/DRY BAY
         Availability:
                Price: $17.05 per kg @ 790 kg
Required Technological
              Changes:

Environmental--Release: ABOUT 4 KG PER INCIDENT
   Combustion Products: 0.6 G HF/G C3H8 BURNED
Toxicity---------Acute: Approx. lethal conc. - 230,000- 255,000 ppm.
              Cardiac: dog; threshold - 26,155 ppm.
         Carcinogenic:
           Mutagenic: Ames assay - negative; human lymphocyte assay -
                      negative; CHO cell assay - negative; Micronucleus
                      assay - negative.
            Pyrolysis:
```

www.ingramcontent.com/pod-product-compliance
Lightning Source LLC
Chambersburg PA
CBHW081854170526
45167CB00007B/3016